ASTRONOMERS' STARS

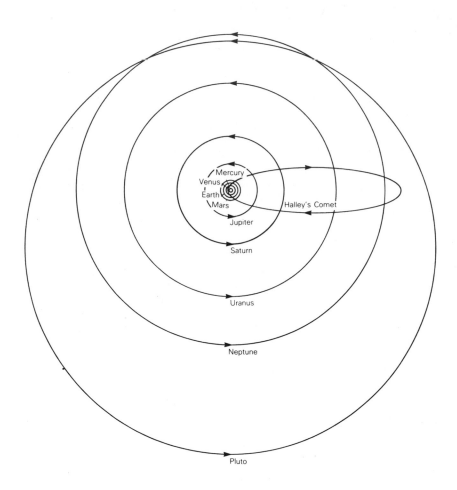

Note: Retrograde Motion of Halley's Comet (opposite direction of travel to planets).

ASTRONOMERS' STARS

Patrick Moore

W. W. NORTON & COMPANY
New York   London

Copyright © 1987 by Hester Myfanwy Woodward
First American Edition, 1989.

Printed in the United States of America.

Library of Congress Cataloging-in-Publication Data

Moore, Patrick.
    Astronomers' stars / Patrick Moore.
        p.   cm.
    Bibliography: p.
    Includes index.
    1. Stars.   I. Title.
QB801.M64 1989
523.8—dc19   88-22376

ISBN 0-393-02663-9

W. W. Norton & Company, Inc., 500 Fifth Avenue, New York, N. Y. 10110
W. W. Norton & Company Ltd., 37 Great Russell Street, London WC1B 3NU

1 2 3 4 5 6 7 8 9 0

CONTENTS

FIGURES

PREFACE AND
ACKNOWLEDGMENTS

Many books on astronomy have been published in recent years, but this one is written in a rather different way. What I have tried to do is to present a survey of 'stellar astronomy' by selecting various stars of particular interest, and telling their stories. They make up a varied collection, from vast Red Giants such as Betelgeux to tiny, super-dense White Dwarfs such as the Companion of Sirius; there are stars which brighten and fade, stars which are members of what may be termed 'family groups', stars which explode, and even objects which look stellar but which prove not to be stars at all.

I have done my best to make the book intelligible to all those who read it, no matter what their age or their previous knowledge. Whether or not I have succeeded must be left to others to judge, but I hope that if you read through these pages you will find at least something to interest you.

My grateful thanks are due to those who have helped me in the preparation of this book – in particular to those who have kindly allowed me to use their photographs; to Paul Doherty, for his usual skilful line drawings; and to the publishers, particularly David Stonestreet.

Patrick Moore
Selsey, 1987

I INTRODUCING THE STARS

Look up into the sky on any dark, clear night, and you will see a great many stars. It is easy to believe that there are millions of them, sometimes so closely crowded together that they are in danger of bumping into each other. This is not true: nobody can ever see more than three thousand stars at any one time, except with optical aid, while on average the stars are millions of millions of miles apart. But it is certainly correct to say that they are numerous: our star-system or Galaxy alone contains at least 100,000 million, and there are many, many galaxies.

What exactly are the stars?

Ancient peoples had no idea. Possibly the stars were holes in the vault of heaven, through which light shone from the cosmic fires beyond; perhaps the stars were tiny lamps fixed to an invisible crystal sphere covering the flat and all-important Earth. There was nothing absurd in such notions to people with no knowledge of science and no clues to the nature of the universe. It was much later that ideas started to change. Slowly, man realized that the Earth is neither flat nor important; the Sun itself is nothing more than a star, and our world moves round it, completing a full journey in one year. Discoveries of this sort were not made easily. The spherical form of the Earth was known long before the time of Christ, but it was not until the work of Isaac Newton, in the seventeenth century, that practically all serious astronomers came to accept that the Earth moves round the Sun.

The distance-scale of the universe was something else that required a great deal of mental adjustment. Ptolemy, last of the great astronomers of Classical times, believed the Sun to be a few millions of miles away; so did Johannes Kepler, who was the first to show the precise way in which the Earth and other planets orbit the Sun – and Kepler was still alive four hundred years ago. It was not until 1672 that the Italian astronomer Giovanni Cassini, by that time director of the new and totally unsuitable observatory at Paris,* established that the Sun's distance is of the order of 86,000,000 miles; we now know that it is just under 93,000,000 miles (150,000,000 kilometres).

But can you really appreciate 93 million miles? Certainly I cannot; I am unable to appreciate even one million miles. Yet the Sun, the ruler and the central body of our Solar System, is a very near neighbour; only the Moon and the three planets Mercury, Venus and Mars are closer to us (neglecting minor objects such as occasional asteroids and comets). The stars are much more remote. They are, indeed, so far away that our conventional measuring units such as the mile and the kilometre become hopelessly inadequate, just as it would be awkward to give the distance between London and New York in inches or centimetres.

The stars are suns, and our Sun, which looks so glorious in our sky, is nothing more than a very average star. It is much less luminous than many of the stars you can see without a telescope: to give just one example, the bright white star, Rigel, in the constellation of Orion, is equal to about 60,000 Suns put together. Its distance is approximately 900 light-years – a term which I must explain at once, because it is all-important in stellar astronomy.

Light does not travel instantaneously. Switch on a torch in a darkened room, and there will be a short interval before the beam reaches the far wall; but I assure you that there is no point in trying to make any measurements, because light flashes along at the staggering rate of 186,000 miles per second (300,000 km per second). This means that light from the Moon, a mere quarter of a million miles away (384,000 km), can reach us in 1¼ seconds; from the Sun, in 8.6 minutes – so that we see the Sun not as it is now, but as it used to be 8.6 minutes ago. In a year, therefore, light can cover rather less than 6 million million miles (9.5 million million km), and this distance is known to astronomers as the light-year. The light from Rigel now reaching our eyes began its journey toward us at the time of William the Conqueror. If some malevolent demon suddenly snatched Rigel out of the cosmos, it would not be until around the year AD 2890 that we would realize that anything untoward had happen to it.

Of course, there are many stars closer than Rigel, but all the stars (apart from the Sun) are light-years from us, so that once we look beyond our own particular part of the Universe our knowledge is

---

*The Paris Observatory was unsuitable because the French king insisted upon its 'looking nice', so that the view of the sky was blocked by various turrets and pieces of architecture which the astronomers did not appreciate in the least. Cassini had to take his telescopes into the grounds and observe from there!

bound to be very out of date. Only in the Solar System can we say that we are seeing things almost as they are 'now'. Round the Sun circle the nine planets – Mercury, Venus, the Earth and Mars close-in; then a wide gap, filled by thousands of dwarf worlds known as asteroids; then the giant planets Jupiter, Saturn, Uranus and Neptune, together with Pluto, a peculiar little rock-and-ice globe whose nature is very uncertain (see frontispiece). Of these, the planets out as far as Saturn were known in ancient times, and there is nothing surprising in this, because they are bright naked-eye objects. Venus, Jupiter and sometimes Mars easily outshine any of the stars, while Saturn is bright enough to be conspicuous, and only Mercury is at all elusive. The three outer planets were found telescopically in modern times: Uranus in 1781, Neptune in 1846 and Pluto as recently as 1930. No doubt there are planets at a still greater distance from the Sun, but even these will be close at hand compared with the stars.

How can you tell a planet from a star? Partly from its appearance; but mainly, at least in pre-telescopic times, from its movement from one night to the next. The stars are so far away that though they are not 'fixed', as was once believed, they move so slowly compared with each other that the patterns or constellations do not change appreciably over many lifetimes. The Great Bear, Orion, the Scorpion and the rest must have been seen by George Washington, King Canute, Jesus Christ, the Trojan warriors and even the mammoth-hunters of the Ice Age. The planets, on the other hand, wander about from one constellation to another. They always keep to certain well-defined limits, and you cannot see them crawling perceptibly against their backgrounds, but if you watch a planet such as Mars on successive nights you can detect its motion easily enough. The very word 'planet' comes from the Greek for 'wanderer'.

In fact, the differences between planets and stars are funda-mental. The planets are members of the Sun's family, and shine by reflected sunlight, whereas the stars are self-luminous. The Moon, which dominates the night sky for long periods each month, is a very minor body; it is our faithful companion, keeping company with us as we journey round the Sun. The Moon's diameter is not much over 2000 miles (around 3500 km), whereas that of the Sun is more than a hundred times greater. The Earth is not unique in having an attendant; the giant planets have whole retinues, numbering at least twenty in the case of Saturn. And in the Solar System we must not forget those ghostly visitors, the comets, whose gleaming heads and long tails sometimes give them a false look of importance. The most famous of all comets, Halley's, last came back to our neighbourhood in 1986; it will be 2061 before it returns once more.

3

The constellations have no real significance. Many people can recognize the seven main stars of the Great Bear, which make up the pattern nicknamed the Plough or the Big Dipper, but not all the seven are at the same distance from us; for example, Alkaid, the 'end' star of the Plough, is much further away than the second star, Mizar. Alkaid is in the background, so to speak, and has absolutely no real link with Mizar. If we happened to be observing from a vantage point in between them, Mizar and Alkaid would be on opposite sides of the sky.

Taking the several thousands of naked-eye stars, you can really make up any patterns you like. The Chinese had their own system; the Egyptians another; the Greeks a third – and it happens to be the Greek system which we use today, though just where it originated is something of a mystery (the island of Crete is one possibility). Ptolemy, who was active around the year AD 150, left us a catalogue of forty-eight constellations, all of which we still accept even though their boundaries have been modified in many cases. The names were drawn partly from mythology (Orion, Perseus, Andromeda and so on) but sometimes from everyday objects (the Triangle and the Altar, for instance). Later astronomers have added constellations of their own, often with up-to-date names such as the Telescope and the Microscope.

Before passing on, let us pause momentarily to dispose of the outdated pseudo-science of astrology, which aims to link the positions of the planets with human character and destiny. For example, certain predictions are made when a planet is 'in' a particular constellation. To take just one case: in June 1986 Mars was seen against the background of stars making up the constellation of Sagittarius, the Archer; but the stars in the Archer are not genuinely associated with each other, and Mars is very much closer than any of them. To claim that Mars was 'in' Sagittarius during June 1986 is about as sensible as claiming that if you hold up a finger and align it with a cloud, your finger is 'in' the cloud. The difference is tremendous. If you represent the Earth-Sun distance by one inch (2½ cm), the nearest star beyond the Solar System will be over four miles (about 6½ km) away.

Look at a bright planet through an adequate telescope, and you will see detail. Mars has dark patches on its reddish surface, together with white polar caps, while Jupiter has cloud belts, and Saturn is surrounded by a magnificent system of rings, made up of icy particles whirling round the planet in the manner of dwarf moons. But no normal telescope will show a star as anything but a point of light, simply because the stars are too far away. Photographically a star will appear as a disc, but this is purely a photographic effect due to the star's brightness. The smaller it appears, either with the camera or with the eye, the better you are seeing it.

The brightnesses, clearly, have a wide range. Sirius is the most brilliant star in the sky (not counting the Sun, of course), and it is a beautiful sight, particularly when low down in the sky and flashing violently; at the other end of the scale there are many stars which are near the limit of naked-eye vision, and are properly seen only with the help of binoculars or a telescope. The apparent brightness of a star is given by its magnitude. The scale here works in the style of a golfer's handicap, with the more brilliant performers having the lower values. Thus a star of magnitude 1 is brighter than a star of magnitude 2; 2 is brighter than 3, and so on down to 6, the faintest stars normally visible with the naked eye. Modern telescopes can go down to as much as magnitude 26. We also have a few stars which are above magnitude 1; Vega in Lyra (the Lyre) is about zero, while Sirius is −1.5. To complete the picture, the brightest planet, Venus, can reach magnitude −4.4, while the magnitude of the Sun is about −26.

The scale is not so casual as might be thought. A star of magnitude 1 is exactly a hundred times brighter than a star of magnitude 6.

One of the constellations in Ptolemy's original list is the Little Bear, always referred to by its Latin name of Ursa Minor simply because Latin is still a convenient universal language. The leader of Ursa Minor is Polaris, or the Pole Star. It is of the second magnitude, and in itself it is unremarkable, though it is much more powerful than the Sun and is nearly 700 light-years away.* It is important to us because it lies practically in the northward direction of the Earth's axis of rotation – so that as the Earth spins, the entire sky seems to move round Polaris (or almost so: the polar point is rather less than one degree of arc away). South of the equator, Polaris never rises. There is a southern pole star, but it is decidedly dim, and is not easy to locate unless you have a good star-map and plenty of patience.

Many of the brighter stars have individual names, such as the familiar Sirius, Rigel and Vega. Actually there are proper names for many of the fainter stars as well, but not many people, even astronomers, would recognize Sarin, Nembus, Alkes or Unukalhai! (For the record, Sarin is Delta Herculis, Nembus is Upsilon Persei, Alkes is Alpha Crateris and Unukalhai is Alpha Serpentis.) Generally speaking, proper names are used only for stars of the first magnitude, plus a few special cases such as Polaris and Mizar. To

*Inevitably there are uncertainties about the distances and luminosities of the more remote stars. In this book, I have followed the authoritative *Cambridge Sky Catalogue 2000*. In many cases other catalogues give rather different values.

confuse things further, stars officially ranked of the first magnitude extend from the brightest of all, Sirius, down to Regulus in Leo, which is only +1.3.

Otherwise, we have recourse to a system which was originally proposed by a German amateur astronomer named Bayer, as long ago as 1603. Bayer, a lawyer by profession, subsequently produced some very peculiar ideas about sky-maps, but his first catalogue of stars was a good one, and his nomenclature was basically sensible. What he did was to take each constellation, and allot the stars in it Greek letters, beginning with the brightest star (Alpha, the first letter of the Greek alphabet) and working through to the last letter, Omega. Thus in Triangulum, the Triangle, the brightest star would be Alpha Trianguli or Alpha of the Triangle; the second brightest star Beta Trianguli, and so on. Inevitably there are many general departures from the general rule and, for example, in Sagittarius (the Archer) the two brightest stars are Epsilon and Sigma, with Alpha and Beta Sagittarii very much in the also-ran category; but it is at least a reasonable guide in most cases. The obvious limitation is that there are only twenty-four Greek letters. John Flamsteed, the second Astronomer Royal, ignored Bayer's system and numbered the stars in each constellation in order of position, so that the brilliant Rigel, Bayer's Beta Orionis, is also Flamsteed's 19 Orionis. But so far as we are concerned at the moment, the proper names and Bayer's letters will do very well, and it may be helpful to give the full Greek alphabet:

Alpha	α	Iota	ι	Rho	ϱ
Beta	β	Kappa	ϰ	Sigma	σ
Gamma	γ	Lambda	λ	Tau	τ
Delta	δ	Mu	μ	Upsilon	υ
Epsilon	ε	Nu	ν	Phi	φ
Zeta	ζ	Xi	ξ	Chi	χ
Eta	η	Omicron	o	Psi	ψ
Theta	θ	Pi	π	Omega	ω

Another obvious fact is that the stars are not all of the same colour. Look at the two leaders of Orion, Rigel and Betelgeux, and you will see this at once, because while Rigel is pure white, Betelgeux is orange-red. Vega in Lyra (the Lyre) is steely-blue; Capella in Auriga (the Charioteer) is yellow, like our Sun, though it is much more luminous. These differences in colour indicate real differences in surface temperature. Blue heat is hotter than white heat, white hotter than yellow, yellow hotter than red, so that of the four stars I have given as examples Vega is the hottest and Betelgeux the coolest.

The colours are striking only for the brightest stars, but this does not mean that the dimmer stars are uniformly white. Far from it. Consider the fourth-magnitude Mu Cephei, in the northern sky. With the naked eye it looks ordinary enough, but as soon as you use a telescope, or even a pair of binoculars, it resembles a glowing coal, and has been aptly nicknamed 'the Garnet Star'. It is much redder than Betelgeux, as we would appreciate at once if it looked brighter. It is, incidentally, much more luminous than Betelgeux, but it is much farther away, at a distance of well over 3000 light-years.

The stars show immensely varied characters. Some are steady in light, others variable; some are single, others double or multiple; some are huge, others tiny. In our system or Galaxy alone there are about a hundred thousand million of them, arranged in a flattened form which resembles the shape of two fried eggs clapped together back to back (Fig. 1); the Sun lies about 30,000 light-years from the centre, and more or less in the main plane. When we look along the 'thickness' of the Galaxy we see many stars in almost the same line of sight, which produces the effect of the lovely Milky Way band which crosses the sky and is a magnificent spectacle on a dark night, when it stretches from horizon to horizon. We have groups or

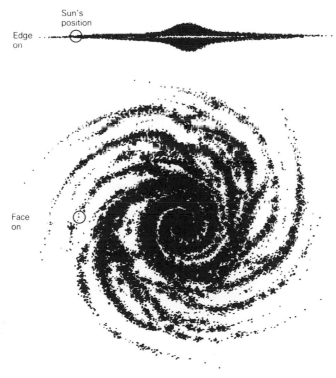

Fig. 1 Shape of the Galaxy

clusters of stars; we have vast gas-and-dust clouds known as nebulæ; we find a tremendous amount of thinly-spread material which we have to study by indirect means. And beyond our Galaxy there are others, most of them so remote that their light takes millions of years to reach us. We can detect objects so far away that we see them as they used to be, long before the Earth itself came into existence.

So let us now turn to some of the individual stars, and decide what they can tell us.

II 61 CYGNI: THE FLYING STAR

Twinkle, twinkle, little star,
How I wonder what you are . . . .

As soon as it became known that the stars are suns, it followed that they must be immensely distant. By everyday standards even our Sun was decidedly remote. Ptolemy, around the year AD 150, gave its distance as 5 million miles. Johannes Kepler, the man who first showed that the planets move round the Sun in elliptical orbits rather than circles, increased Ptolemy's value to 14 million miles in 1618, when he published the last of his classic Laws of Planetary Motion. Then, in 1672, the Italian astronomer, Giovanni Cassini, working in France, gave the Earth-Sun distance as 86 million miles, which is not far short of the real distance of just under 93 million miles (150 million km).

Cassini's method was of particular interest, because it was exactly the same, in principle, as the method used by the early investigators to try to find out the distances of the stars. He depended upon parallax, which is easy to explain by a very simple experiment.

First, close one eye and hold up a finger at arm's-length, lining it up with a relatively distant object such as a tree in the garden (Fig. 2a). Now, without moving your finger or your aim, use the other eye. Your finger will no longer be lined up with the tree, because you are looking at it from a slightly different direction; your two eyes are not in the same place. The distance between your eyes is the baseline, and the angle P is a measure of the parallax. If you know the length of the baseline, and also the angle at P, you can solve the whole triangle, and find out how far away your finger is.

Of course this depends upon assuming that the tree shows no parallax at all, but if the tree is sufficiently far away its own parallax shift can be ignored. The principle can next be extended to

9

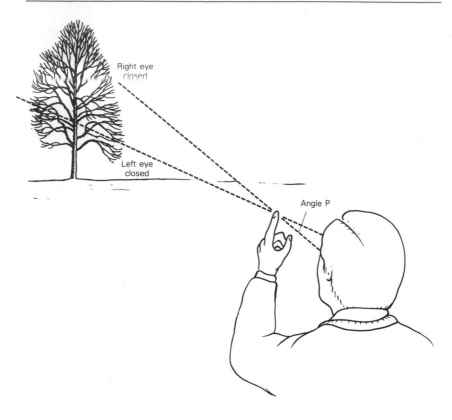

Fig. 2(a) Parallax demonstration

measuring the distance of some inaccessible object, such as a mountain top. The surveyor marks out two observation points, suitably separated, and measures the angle of the peak from each. He can then solve the triangle and obtain the answer he wants, which is much easier than undertaking a long and dangerous climb. (A slight correction has to be made for altitude, of course, but this is easy enough.)

Now, let us come back to Cassini – who seems, incidentally, to have been a somewhat unpleasant character, but whose contributions to astronomy were very great. Earlier in the seventeenth century Kepler had drawn up his famous laws, which provided an accurate scale model of the Solar System. What he lacked was a knowledge of one actual distance. For example, suppose we could find out the real distance of the planet Mars. Kepler's Laws tell us that the distance between Mars and the Sun is 1.524 times that between the Earth and the Sun. Therefore, given the real figure for Mars, that of the Earth can be worked out by simple arithmetic.

This is what Cassini set out to do. Mars was well placed for observation in 1672, and Cassini sent his assistant, Jean Richer, to Cayenne, in South America, while he himself remained in Paris. Both observers measured the position of Mars against the background stars; this time the baseline was over 6000 miles, the distance between Paris and Cayenne. From the results, Cassini found the parallax of Mars, calculated its distance, and gave a value for the length of the astronomical unit (Earth-Sun distance) which was only about 7 million miles too small.

Cassini's subsequent behaviour was not to his credit. Richer's work had been extremely good, and Cassini was jealous. Accordingly he banished his unlucky assistant to work on some military fortification projects in a remote province, and little more was heard of him. Moreover, Cassini was ultra-conservative, and never accepted the theory that the Earth moves round the Sun. But at least he had shown that the parallax method works, and the next step was to apply it to the stars.

Quite obviously it was necessary to use a baseline much longer than the diameter of the Earth. Mars, only a few tens of millions of

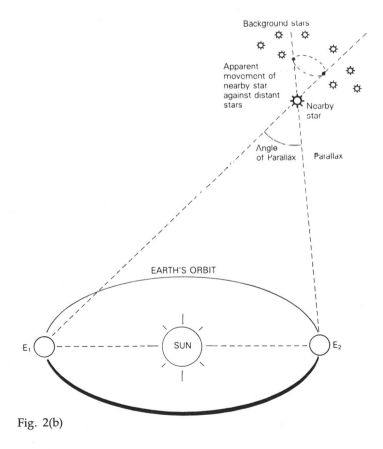

Fig. 2(b)

miles away from us, is so near on the cosmic scale that the stars can be regarded as infinitely remote, just as in our original demonstration the garden tree can be regarded as infinitely remote. But for the stars, the first essential is to select one which is closer than its neighbours; and as they all seem, at first glance, to be equally far away – there are no 3D effects in space – this is not easy.

The obvious baseline to use was the diameter of the Earth's orbit (Fig. 2b). As a start, select your relatively nearby star, (Just how to tackle this I will discuss in a few moments). Assume that the background stars are so far off that they show no measurable parallax. Now observe the nearby star twice, once in January (E1) and then in July (E2), by which time the Earth has moved from one side of the Sun to the other. The baseline E1–E2 is now 186 million miles (300 million km), the full diameter of our orbit. The nearby star will seem to have moved perceptibly. Again, we can find the angle at P, and then solve the triangle in the usual way. The main snag is that the angle at P is bound to be very small indeed.

James Bradley, who succeeded Edmond Halley as Astronomer Royal at Greenwich in 1742, decided to take up the problem. In fact he did so long before he achieved fame – his work dated from 1725, and was carried out not at Greenwich but at Kew, where a friend of his named Molyneux had a private observatory. The star he selected was Gamma Draconis, in the constellation of the Dragon, which is of the second magnitude, about the same as Polaris. From southern England, Gamma Draconis passes directly overhead once every twenty-four hours, and Bradley reasoned that it would be very easy to measure, because it could be observed with a telescope which would point straight upward and would not have to be manoeuvrable to any marked extent.

He and Molyneux observed regularly for a year, and found something very strange indeed. Gamma Draconis shifted, but not because of parallax; it appeared to describe a small circle in the sky relative to the stars nearby in the sky. Bradley was puzzled. Then, according to tradition, he went for a boating trip on the Thames and hit upon the solution. He saw that when the boat altered direction, the vane on the mast-head shifted slightly even though the wind direction was the same as before. The same principle could explain the behaviour of Gamma Draconis. The Earth is moving round the Sun, so that its direction is changing all the time, but the light from the star is coming in at a constant direction. If the Earth is represented by the boat, and the starlight by the incoming wind, it is obvious that there will always be an apparent shift of the star toward the direction in which the Earth is moving at that particular moment. An even more common example is that of a walker who is carrying an umbrella and is caught in a rainstorm. If he wants to keep dry, he will not hold the umbrella vertically; he will slant it forward, in the direction in which he is walking.

Bradley had failed to measure the distance of a star, but he had discovered what we now call the aberration of light (Fig. 3). It was, incidentally, the first practical observational proof that the Earth is moving in space.

Gamma Draconis was a bad choice in any event. We now know that its distance is over 100 light-years, so that its parallax would have been much too small to have been measured with the instruments used by Bradley and Molyneux. All the same, it was a brave attempt.

The next major character in the story is William Herschel, who may well be regarded as the greatest astronomical observer of all time. He was Hanoverian by birth, but came to England when still a young man, and spent the rest of his life there (thereby becoming 'William' instead of 'Wilhelm'; he was knighted late in his career). He became organist at the famous Octagon Chapel in Bath, but gradually his interest in astronomy began to overshadow everything else. He made reflecting telescopes which were the best of their time

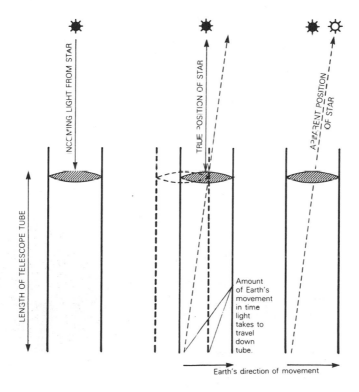

Fig. 3 Principle of the aberration of light

13

– his largest, completed in 1789, had a metal mirror 49 inches in diameter, still on view at the Old Royal Observatory in Greenwich Park – and his skill was almost uncanny. In 1781 he discovered a new planet, the one we call Uranus, and his reputation became worldwide. A grant from King George III of England and Hanover enabled him to give up music as a career, and, helped by his faithful sister, Caroline, he settled down to almost forty years of careful, systematic work. He determined the rotation period of Mars to an accuracy of a few seconds; he found the direction in the sky toward which the Sun is moving, carrying the Earth and the other planets with it; he discovered thousands of new double stars, clusters and nebulæ, and he was the first to give an accurate picture of the shape of the Galaxy. There was seemingly nothing he could not do – except measure the distance of a star.

Herschel realized that the parallax method was the only one available, and at least he had the advantage of being able to use telescopes much more powerful than Bradley's. To offset the problem of identifying a relatively close star, he proposed to make use of doubles.

There are many double stars in the sky; the first to be discovered telescopically was Mizar, in the Great Bear, about which I will have more to say in the next chapter. Another famous double is Castor, in Gemini, the senior but fainter member of the Twins; it is half a magnitude dimmer than its neighbour Pollux. From all accounts it was Cassini who, in 1678, first realized that Castor is double, and it was certainly observed as such by Bradley in the mid-eighteenth century. The two members of the pair (designated A and B) are too close together to be seen separately with the naked eye. Their magnitudes are 2.0 and 2.8 respectively, and at the present time are separated by about 2 seconds of arc. This means that the pair can be split with any reasonable telescope of 'amateur size'.

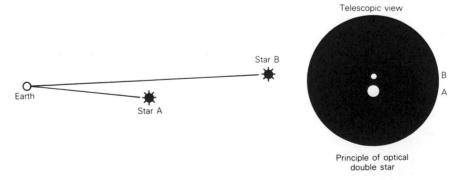

Fig. 4 Optical double principle

14

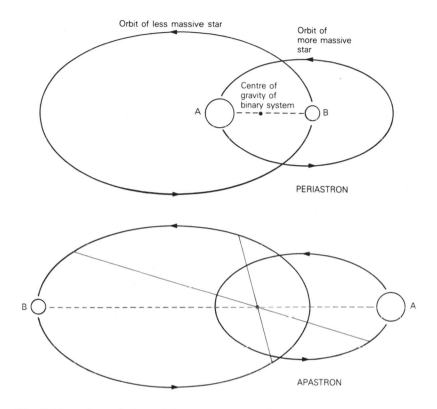

Fig. 5 Mutual revolution of the components of a binary

Presumably, thought Herschel, we must be dealing with a line-of-sight effect. In Fig. 4 we have two stars, A and B, which are at different distances from us; A is in the foreground. This means that if we use the Earth's orbit as a baseline, and assume that B is remote enough to show no detectable parallax, we can find out the parallax of A. To make this clear assume that, in January, A is on one side of B, while in July it is on the other side. Of course the shift is extremely slight, but the principle is sound. Herschel set out to measure several pairs of stars, including Castor.

Like Bradley before him, he made an unexpected discovery. He found no parallax shifts, but he did find that in some cases the components of a double star were moving round each other, as the two bells of a dumbbell will do if you twist them by the bar joining them (Fig. 5). This was true of Castor. A and B are moving round their common centre of gravity in a period which was determined by W. Rabe, in 1965, as being as long as 420 years. The apparent separation is less now than it used to be a century ago, when it was over 6 seconds of arc. The two components are more 'lined up',

15

though they are now very gradually starting to separate again as seen from Earth.

Herschel failed to detect the parallax of any star only because his measuring instruments, good though they were, were still too crude. But he had at least made a fundamental discovery during his search; he may not have been the first to suggest the existence of physically associated or binary systems (probably he was anticipated in 1767 by the Rev. John Michell), but he was the first to prove it observationally. It was in 1803 that he made his announcement about the binary nature of Castor.

In fact, Castor is more complicated than Herschel had supposed. Each main component is itself a close binary, and there is a third, fainter member of the system, which is also made up of two, so that Castor is a sextuple system – a sort of stellar family. The distance is 46 light-years, which means that its parallax is much too small for Herschel to have measured. So let us turn now to the star which really did give the key to the whole problem: 61 Cygni, in the constellation of the Swan.

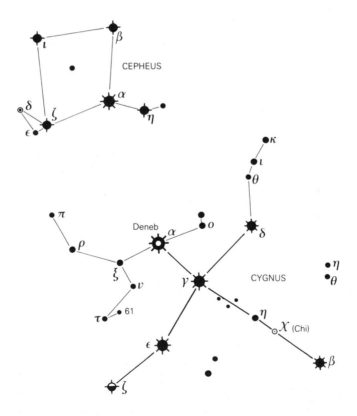

Fig. 6 Position of 61 Cygni

There is nothing at all striking about 61 Cygni. It is of the fifth magnitude, so that it is easy enough to see with the naked eye on a dark, clear night, but you will have to look carefully for it (Fig. 6). Cygnus itself – often nicknamed the Northern Cross – is a splendid constellation; its five main stars make up an X-pattern, and one of them, Deneb, is of the first magnitude. Deneb is a cosmic searchlight, perhaps 70,000 times more powerful than the Sun; it never actually sets over England, though at its lowest it almost grazes the horizon. Cygnus is high in the evening sky all through the summer. Deneb makes up a large triangle with two other bright stars, Vega in Lyra and Altair in Aquila (the Eagle). Many years ago, in a BBC *Sky at Night* television programme, I gave the three the nickname of 'the Summer Triangle', and this term has come into everyday use, though it is quite unofficial – and is inapplicable to the southern hemisphere, where the three stars are always rather low down and are visible in the evenings only during the southern winter. From parts of New Zealand, Vega and Deneb do not rise at all.

The faintest star of the cross of Cygnus, Albireo or Beta Cygni, lies some way off the line joining Vega to Altair; it is fainter than the other stars in the cross (despite being dignified by the second letter of the Greek alphabet), but is a glorious telescopic double, with a golden-yellow primary and a blue companion. To locate 61 Cygni, first find Gamma or Sadr, the central star of the X, and then Epsilon, of magnitude 2.5. Tau Cygni (magnitude 3.7) makes up a slightly distorted quadrilateral with Gamma, Epsilon and Deneb. 61 Cygni is close to Tau, and in the same binocular field.

Use a telescope, and you will find that 61 Cygni is an easy double. The components are of magnitudes 5.5 and 6.3, and the separation is 27 seconds of arc. Both components are orange, and of spectral type K. Observations of the relative movements of the two since 1830, when the first measurements were made by F. G. W. Struve from the Dorpat Observatory in Estonia, have led to an estimated revolution period of around 650 years. But as early as 1792, 61 Cygni had attracted attention for quite another reason. It was found to have an abnormally large individual or 'proper' motion.

Originally the stars were called 'fixed stars', to distinguish them from the wandering stars or planets; it was, rather naturally, believed that they kept exactly to the same relative positions. It was Edmond Halley who first showed that several brilliant stars, including Sirius, had shifted perceptibly since Classical times. The proper motions were very slight – in the case of Sirius, only 1.3 seconds of arc per year – but they were detectable. There could no longer be any possibility that the stars were genuinely fixed in space.

17

Very few stars have proper motions of more than one second of arc per year, and this is not much. Consider a circle which is 2 centimetres across. View it from a distance of 4 kilometres or 2½ miles, and it will subtend an angle of precisely one second of arc. Any star with a proper motion of more than 3 or 4 seconds of arc per year must be assumed to be relatively close, and in 1792 61 Cygni was found to have the high proper motion of 5.2 seconds of arc. Also it was a wide binary, and the spectra indicated that both stars were feeble red dwarfs. We now know that the brighter component (A) has less than 7 per cent of the luminosity of the Sun; the fainter (B) less than 4 per cent. They are true cosmic glow-worms.

In every way, then, 61 Cygni appeared to be a strong candidate for exceptional closeness. Rapid proper motion (it was even nicknamed 'the Flying Star'), low luminosity, wide separation of its two components . . . everything fitted. And in 1838 it was the star selected by the German astronomer, Friedrich Wilhelm Bessel, for a new attack on the problem of stellar distances.

Bessel was born in 1784. At the age of fifteen he began work as an apprentice in a business firm, but commerce did not appeal to him at all, and he cast around for a career which would really interest him. He studied geography, mathematics and navigation; he became a good linguist; in 1804 he dared to write a paper on an astronomical subject (it concerned the orbit of Halley's Comet) and sent it to Dr Heinrich Olbers, who was a famous amateur astronomer. Olbers, a leading medical researcher, was known to be exceptionally pleasant and helpful, and in his treatment of Bessel this side of his nature was fully shown. He arranged for the paper to be published, and then introduced Bessel to Johann Hieronymus Schröter, chief magistrate of Lilienthal (near Bremen), who had a private observatory and was the real founder of the study of the Moon's surface.

Schröter – like Olbers, friendly and helpful – was duly impressed, and invited Bessel to become his assistant. Bessel was only too pleased to accept, and, thankfully, turned his back forever on the world of commerce.

He stayed at Lilienthal for four years, but then, in 1808, the Prussian government decided to build a new observatory at Königsberg, and looked round for a young, skilful and enthusiastic director. Bessel appeared to be ideal. He was invited to go there, and in 1810 became the first director of the first major German observatory. He remained there for the rest of his life, and carried through a series of brilliant research programmes as well as proving himself to be a highly capable administrator. It was a tribute to him when Olbers commented that his own most important contribution to science was his recognition of Bessel's outstanding ability!

Bessel was a superbly accurate observer. For example, he corrected observations for what is now termed 'personal equation', characteristic of the observer concerned; thus if a timing is to be made of the transit of a star across the wire of a measuring device such as a micrometer, some observers will make the transit slightly too early and others will be slightly too late – but the amount of personal error will be consistent, so that it can be taken into account.* Bessel also did everything possible to eliminate errors due to the actual instruments. All the corrections were small, but when dealing with very tiny quantities they were vitally important; and this, of course, was before photography had been developed to take over from sheer visual observation at the eye-end of a telescope.

Star distances had defeated all observers up to that time, even Herschel. But Bessel's equipment at Königsberg, including a new type of micrometer for accurate measurements of star positions – combined with his own skill, and the fact that his telescopes were better mounted than Herschel's – indicated that success might be possible. So Bessel began work, taking 61 Cygni as his target, and within a year he had found a parallax shift which gave a distance of around 10 light-years. As the real distance is 11.2 light-years, he was very close to the truth. For the first time, the cosmic distance-scale had been found; Olbers, who received Bessel's results on his eightieth birthday, commented that the measurement of 61 Cygni had 'put our ideas about the universe on a sound basis'.

The instrument which Bessel used was known as a heliometer. Its main light-collector was an object-glass which had been cut in half, and therefore produced a double image of the image under study. The halves were then slid until the image became single, and the amount of movement needed was a key to the apparent movement of the object involved. Bessel's heliometer had been made by Josef Fraunhofer, the best optical worker of the time, and the original intention had been to use it for measuring the separations of double stars, but Bessel adapted it during his studies of 61 Cygni.

Bessel announced his results in December 1838. He then dismantled the heliometer, overhauled it, and made a second series of over four hundred measurements, ending in March 1840. The results were practically the same as before. So to Bessel goes the honour of being the first man to measure the distance of a star – but it might so easily have been otherwise.

The first major observatory in the southern hemisphere was at the Cape of Good Hope. In 1832 Thomas Henderson, from Edinburgh,

---

*During the war I was an RAF navigator, and in those far-off days we used 'astro', which meant measuring the altitudes of navigation stars by a sextant. After weeks of experimenting I found that all my own estimates were slightly too high. When I allowed for this, they were accurate – at least, I hope so!

was appointed director in succession to the first holder of the office, the Rev. Fearon Fallows. Henderson frankly hated the place; he referred to it as a 'dismal swamp', and he stayed there for only little over a year, after which he retired gratefully to his native Scotland. But in that brief period he carried out some important work, and in particular he made a series of measurements of the third brightest star in the sky, Alpha Centauri, which had a large proper motion (3.7 seconds of arc per year) and was also a wide binary with a revolution period of 80 years. Alpha Centauri is one of the two Pointers to the Southern Cross (Fig. 7). Neither it nor the Cross can be seen from anywhere in Europe, but to South Africans, Australians and New Zealanders it is as familiar as the Great Bear and Orion are to Englishmen.

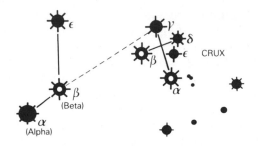

Fig. 7 Position of Alpha Centauri, with the Southern Cross

Henderson's observations of Alpha Centauri, made from the Cape, began in April 1832, and on his return to Scotland he started to work them out. It soon became evident that there was a considerable parallax effect, but Henderson delayed making the final calculations, and he was also anxious to obtain some extra measurements, which could not be done from Scotland simply because Alpha Centauri never rises above the horizon. Finally he was able to give a value for the parallax of 0.75 seconds of arc, corresponding to a real distance of just over 4 light-years. He made his announcement on 9 January 1839, just two months after Bessel had made known his results for 61 Cygni, and so he lost the honour of priority.

Actually his work was rather easier to execute than Bessel's, because the parallax was greater. Alpha Centauri, as we have noted, is a wide binary; one component is slightly more luminous than the Sun, the other rather less. The third member of the system, Proxima Centauri, is a feeble red dwarf which is slightly closer to us, and at its distance of 4.2 light-years it is in fact the very nearest star to the Sun.

Also in the 1835–8 period, F. G. W. Struve at Dorpat, using equipment made by Fraunhofer, had been trying to obtain a parallax for Vega. He found a displacement, but as the value he gave was much too large, and as his announcement was delayed in any case, it cannot be said that his work rivalled that of Henderson – still less that of Bessel.

So 61 Cygni retains its place in the history of astronomy, and of the naked-eye stars only three (Sirius, Procyon in the Little Dog, and the much fainter Epsilon Eridani, in the River) are closer. It has also proved to be a very interesting system in its own right, because the fainter component seems to be associated with an invisible companion which may be too low in mass to be a star, and could possibly be a planet. It would be fascinating to speculate about the view from such a world; there would be two red suns in the sky, and during darkness the constellations would not appear very different from ours, but our Sun would be reduced to a modest star of about the fourth magnitude.

The 1838 success may be regarded as the climax of Bessel's career, but he undertook many other investigations, notably with regard to Sirius and Procyon, about which I will have more to say later. He also studied the movements of Uranus, which was then the farthest-known planet from the Sun. But for his illness, followed by his premature death at Königsberg in 1846, he might have been the first to locate the planet we now call Neptune. This was not to be, but Friedrich Bessel will never be forgotten. It was he who gave the first conclusive proof of the distance-scale of the universe.

III MIZAR: THE HORSE AND HIS RIDER

Of all the constellations, the three best known are undoubtedly Ursa Major, Orion and Crux Australis. Of these, one – Crux Australis, the Southern Cross – is too far south to be seen from Europe or virtually the whole of the mainland United States (though it is easy enough from Hawaii). Orion, the celestial hunter, is crossed by the equator of the sky, so that it can be seen from every inhabited continent. Ursa Major, the Great Bear, is in the far north. From Britain it is circumpolar, so that it is always on view whenever the sky is sufficiently clear and dark; from countries such as South Africa and Australia it is inconveniently low, and from most of New Zealand it never rises at all. Its seven chief stars are arranged in the pattern which is known variously as the Plough, the Big Dipper and King Charles' Wain (Fig. 8).

To Englishmen (such as myself) the night sky seems barren without the Great Bear, but it is interesting to see both it and the Southern Cross on view at the same time; this can be done, for example, from Indonesia, which is very close to the equator, so that the celestial poles are at opposite horizons and the celestial equator passes overhead. Of the two groups the Cross is much the smaller, and is in fact the smallest of all the eighty-eight recognized constellations, but it has two first-magnitude stars and one only just below, while the brightest stars of the Bear are not much above the second magnitude.

The stars of the Plough pattern are, in order of position, Alpha (Dubhe), Beta (Merak), Gamma (Phad), Delta (Megrez), Epsilon (Alioth), Zeta (Mizar) and Eta (Alkaid or Benetnasch). Astronomers seldom use the proper names except in the case of the most celebrated of the Plough stars, Mizar. The name, like so many others, is Arabic, and means a waistband or girdle.

With the naked eye Mizar appears as a star of magnitude 2.1,

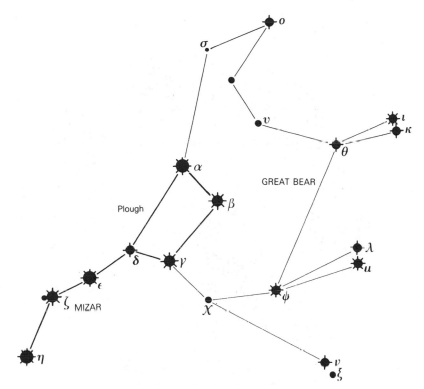

Fig. 8 Great Bear, showing Mizar

marginally fainter than Polaris. What marks it out at once is the presence of a companion star close beside it (Fig. 9). The companion is named Alcor, from the Arabic 'Al-jat', a rider – hence the old nickname of the pair as the Horse and his Rider. Alcor's number, given by John Flamsteed during the compilation of his star-catalogue, is 80, so that its official designation is 80 Ursæ Majoris. The magnitude is exactly 4.

And this brings us on to our first problem. Under even reasonable conditions there is absolutely no difficulty in seeing Alcor with the naked eye. It is 11.8 minutes of arc away from Mizar, so that it is in no sense overpowered (as, for instance, the large satellites of Jupiter are by the brilliance of the planet itself). Even thin mist is barely sufficient to hide Alcor. I once carried out a test with a party of forty people who knew nothing about astronomy and cared less. The sky was reasonably dark; I indicated the Plough, and asked whether any of its stars showed an unusual feature. Every member of the party pointed out Alcor.

And yet the Arabs of a thousand years ago made it quite clear that they regarded Alcor as a naked-eye test! It was sometimes termed

23

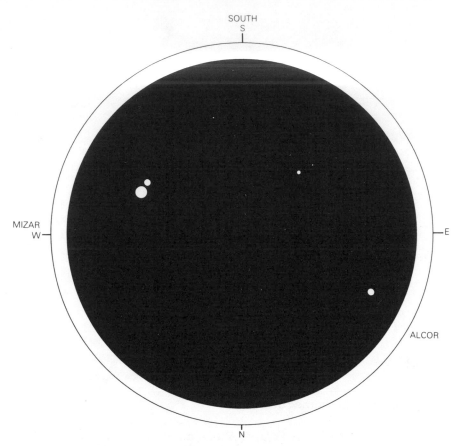

Fig. 9 Mizar through the telescope

Suhā ('The Forgotten One') because it could be seen only by a keen-sighted observer. In the thirteenth century a Persian writer, Al-Kazwini, made the categorical statement that 'people tested their eyesight by this star', while a hundred years later it was referred to as Al-Saduk, 'The Test' (by the Arab writer Al-Firuzabadi). In general the Arabs had excellent sight, and the star catalogues which they produced were much better than those of the Greeks. From Arab countries the Plough rises high, and there was much less light pollution and industrial haze than there is now. So what is the answer?

Alcor has an A5-type spectrum (see p. 39). It is about twelve times as luminous as the Sun, and there is nothing to indicate that it is likely to show any marked change in brightness over a period as short as a thousand years. It is most unlikely that there has been any real alteration. It may be significant that there is an associated minor mystery concerning Megrez, the faintest of the Plough stars, which according to the Greeks was as bright as the rest, though it is now a magnitude dimmer. Megrez also seems to be a perfectly normal, stable star, and it is probable that we are dealing with an error in either observation or interpretation. This might apply equally to the Arabs' description of Alcor.

The only other possible answer is that the Arab 'test star' was not Alcor at all, and this cannot be completely ruled out. Use any small telescope, and you will see not only Mizar (which is itself double) and Alcor, but also a much fainter star of the eighth magnitude. Apparently this star was first noted in 1681 by Georg Christoph Eimmart, who lived in Nürnberg (Nuremberg) and is best remembered for compiling a small map of the Moon (nowadays a lunar crater is named after him). Eimmart's observation aroused no particular interest, and the next report came in 1723, when a German observer whose name has not been preserved saw it again. He believed it to be either a new star or, more probably, a planet (!), which shows that his knowledge of astronomy can hardly have been very great. He named it Sidus Ludovicianum, in honour of Ludwig V, Landgrave of Hesse.

Ludwig's star lies roughly half-way between Alcor and Mizar; if it were brighter – say of the fourth magnitude, equal to Alcor – it would indeed be a test for keen eyes. Whether or not it is variable does not seem to have been established, but the evidence is against anything of the sort, because the Mizar group has been photographed so often over the past hundred and fifty years that any increase in Sidus Ludovicianum could not have passed unnoticed. Therefore the mystery remains – but if there ever is a sudden, temporary flare-up of Ludwig's star we will at least know what the Arabs meant.

Mizar itself has the distinction of being the first double star to be discovered telescopically. The astronomer responsible was a Jesuit, Joannes Baptista Riccioli, who seems to have been a curious mixture. He became professor of philosophy, theology and astronomy in Bologna, and is well remembered for producing a map of the Moon, in 1651, in which he allotted names to the main craters and other features – names which are still in use today. (Immodestly, he named a particularly large walled plain after himself, and an even larger one after his pupil, Grimaldi, upon whose work the map was mainly based.) Riccioli was a good observer, and yet he could never bring himself to believe that the Earth moves round the Sun; he preferred a kind of hybrid system according to which the planets orbited the Sun, but the Sun orbited the Earth.

Telescopes, first used for sky-watching around 1609–10, had come into vogue, and Riccioli of course had one. He looked at Mizar and discovered that instead of one star he was seeing two – quite apart

25

from Alcor (and no doubt Sidus Ludovicianum also, though in fact he did not mention it). Of the two Mizars, one was decidedly brighter than the other; the modern magnitude values are 2.3 for the primary (Mizar A) and 3.9 for the secondary (Mizar B). With the naked eye they merge, increasing the combined magnitude to 2.1. The separation between the two is 14.5 seconds of arc, so that there is no chance at all of resolving the pair with the naked eye or even binoculars; we cannot explain the Arab problem that way.

Mizar is a very easy pair, and any small modern telescope will split it, but, curiously, there are reports that some eighteenth-century astronomers failed to do so; such was the experience of the French astronomer, Honoré Flaugergues, before 1787, though he saw it later. The famous comet-hunter, Pierre Méchain, was similarly unsuccessful, which seems to indicate that his telescope was not really good enough for anything except searching for comets. But again there is no question of real change; G. Kirch saw Mizar double in 1700, and measurements of it have been made regularly since Bradley's work around 1755. Much later, in 1857, Mizar became the first double star to be photographed – by G. P. Bond, one of the earliest American astronomers, who carried out his pioneer work from the then newly-founded observatory at Harvard College.

As we have seen, there are two kinds of double stars: optical pairs, and binaries. Rather surprisingly, binaries are much the more common, and from the outset there was little doubt that the two Mizars were genuinely associated. Yet they are a long way apart, so that if they are moving round their common centre of gravity their revolution period must amount to many thousands of years. The only way to find out is to measure what is termed the position angle, or direction of one star referred to the other (Fig. 10). Conventionally, the brighter star is taken as the standard, and the position angle of the secondary is measured, from north (0 degrees) through east (90), south (180), west (270) and north (back to 0, or 360 degrees). In 1755 Bradley made the position angle 143°. It is now a little over 150°, so that there has been a slight but perceptible change, assuming that Bradley's work was up to its usual high standard. Trying to obtain a reliable period is a hopeless task; all we can really say is that the two Mizars have a real relationship, and that they are moving together in space.

The next step is to find out the distance of the pair. The proper motion is very slight, and amounts to only about 0.12 seconds of arc annually. This means that the parallax is so tiny that it is hard to measure, and estimates of Mizar's distance are not in complete agreement. For a long time the general value was taken to be 88 light-years, but in the new Cambridge sky catalogue this is reduced

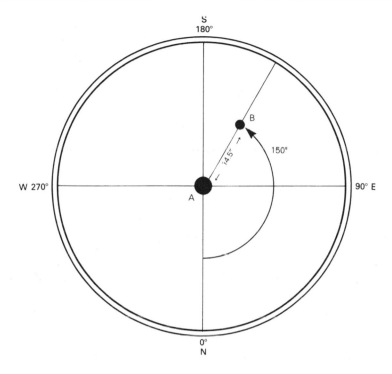

Fig. 10 Position angle

to only 59 light-years – an extra indication that once we move out beyond the immediate neighbourhood of the Sun, all our measurements are subject to very marked uncertainties.

Both the Mizars are white, with A-type spectra, so that their surfaces are hotter than that of the Sun. The primary is thought to be about seventy times as luminous as the Sun, the secondary about thirteen times, so that Mizar B is about as powerful as Alcor. The present distance between Mizar A and Mizar B is around 35,000 million miles, which is about five times the diameter of Pluto's orbit in the Solar System, and is about one-twentieth of a light-year.

But is Alcor also a member of the group? All logical reasoning says 'yes'. Its direction of movement in space is much the same as that of the Mizar pair, and most estimates give the real distance between Mizar and Alcor as no more than a quarter of a light-year. Yet in the Cambridge catalogue the difference in distance is more than 20 light-years, with Alcor the more remote. On the whole it seems that the older value is the more likely; it is hard to believe that the two Mizars are totally distinct from Alcor, though Ludwig's star does indeed lie in the background, and is merely seen in much the same direction from Earth.

27

It is also worth noting that of the seven stars in the Plough, five make up what is termed a 'moving cluster', and are travelling in the same direction at much the same rate round the centre of the Galaxy, so that presumably they have a common origin. The two exceptions are Alkaid (much the most luminous of the seven) and Dubhe (which, unlike the rest, is an old orange-red star). Over a sufficiently long period the Plough shape will become distorted because of the different movements of Alkaid and Dubhe, but the proper motions are so small that they are inappreciable over many lifetimes from the viewpoint of the naked-eye observer. If you want to see any obvious change in the Plough, you will have to come back several tens of thousands of years hence.

Mizar has yet another distinction. The primary, A, was the first star to be identified as a spectroscopic binary – an episode which is of tremendous importance, and is worth describing in rather more detail.

What we normally call 'white' light is in fact a medley of all the colours of the rainbow. The colour of light depends upon its wavelength, and light is a wave motion, so that the principle is much the same as that of sound waves in the air or, for that matter, waves on the surface of the sea. The wavelength is the distance between one wave crest and the next. In what we call the 'electromagnetic spectrum' (Fig. 11), only a small part makes up the range of visible light. In this section red light has the longest wavelength; then come orange, yellow, green, blue and violet. If the wavelength is longer than that of red light, it does not affect our eyes, and we cannot see it; we have come to the infra-red range (most people are familiar with the infra-red lamps used in hospitals). If you want to detect it, all you have to do is switch on an electric fire: you will feel the infra-red, in the form of heat, well before the bars become hot enough to glow. If the wavelength is longer still, we come to the region of radio waves. At the other end of the

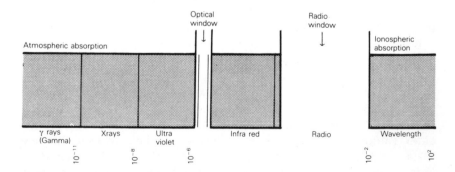

Fig. 11 The Electromagnetic Spectrum

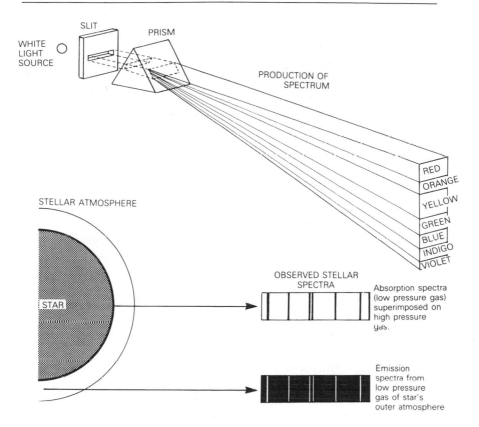

Fig. 12 Production of a stellar spectrum

spectrum, on the short side of violet light we have ultra-violet, X-rays, and finally the extremely short gamma rays.

Of course, even red light has a wavelength which is very short by everyday standards, and amounts to a tiny fraction of a millimetre. But when light is passed through a glass prism, or some equivalent device, the wavelengths are sorted out to produce the colours of the rainbow (Fig. 12). This was first done, in principle, by Isaac Newton in 1666, when he was isolated in his Lincolnshire cottage because Cambridge University had been closed during the Plague.

If you pass the light from an incandescent solid, liquid or high-pressure gas through a prism, you will obtain a rainbow from red through to violet; this is termed a continuous spectrum. Analyse the light from a low-pressure gas, and the result is quite different. There is no rainbow, but only a series of bright, isolated coloured lines, forming what is termed an emission spectrum. Each of these lines is due to some particular element or group of elements, and cannot be duplicated. Thus if you see two bright yellow lines, in a definite

29

position, they can be due only to sodium, one of the two elements making up common salt. Nothing else can produce them. Some spectra are immensely complicated; for instance, that of iron contains many thousands of lines.

Now consider a star such as the Sun. The bright surface is made up of high-pressure gas, and yields a continuous spectrum. Overlying this surface is a region of low-pressure gas, which would normally produce an emission spectrum. But the lines which would otherwise be bright are seen against the rainbow background, and are 'reversed', so that they seem dark; we now have what is termed an absorption spectrum. The positions and intensities of the lines are not affected, so that they can still be identified. The yellow part of the Sun's spectrum shows two dark lines in the precise positions where the bright yellow lines of sodium would be expected to be; therefore, we can tell that there is sodium in the Sun.

With the Sun, of course, there is always plenty of light to spare. Not so with the stars; to split up starlight into a spectrum means using a spectroscope combined with a powerful telescope. Today, virtually all such work is carried out by photography or electronic aids. The days of visual observation at the eye-end of a telescope are to all intents and purposes over.

Using a spectroscope, then, we can find out what elements are present in the stars. But we can do more; for example, we can find out the ways in which the stars are moving.

Listen to a passing police car or ambulance which is sounding its siren, and you will find that the note is high-pitched when the car is approaching you, lower-pitched after it has started to move away. When the car is approaching, more sound-waves per second reach your ear than would be the case if the car were motionless; the wavelength of the sound is effectively shortened, causing the note to be raised. After closest approach, fewer sound-waves per second reach you, so that the wavelength is apparently lengthened and the note drops. Much the same thing happens with light. An approaching object will be slightly 'too blue', while a receding object will be 'too red'. This has been known for well over a hundred years; it is known as the Doppler effect, after the Austrian physicist who first drew attention to the principle.

So far as a star is concerned, the actual colour change is too slight to be noticed, but the effect of movement shows up in the spectrum. With an approaching star, all the dark (absorption) lines are shifted over to the short-wave or blue end of the rainbow; if the star is receding, the shift is to the red – and the amount of the shift gives a key to the rate at which the star is moving toward or away from us. Thus with Mizar, the Doppler shift indicates that the pair is approaching us at the rate of 5½ miles per second (about 8 km per second). This may seem a tremendous speed, but it does not mean that Mizar will eventually collide with us; by the time it could reach our part of the Galaxy, we would be somewhere else! These so-called radial velocities of the stars are bound up with the overall rotation of the Galaxy around the centre of the system, about 30,000 light-years away from us.

The first careful studies of the spectra of the stars were made in the second half of the nineteenth century. One of the great pioneers was Edward C. Pickering, who was born in Boston in 1846. He trained as a mathematician and physicist, and then, in 1876, he was appointed director of the Harvard College Observatory, a post which he held with great distinction for the next forty-two years. It was he who was the prime mover in compiling a great star catalogue which was regarded as the standard for many years, and he carried out important researches in all branches of astronomy. (His brother William was also an eminent astronomer, though more concerned with the Moon and planets than with the stars.)

In 1889 Edward Pickering began a study of the spectrum of Mizar A, the brighter component of the pair. He made a very curious discovery. Sometimes the absorption lines were single, and the spectrum appeared quite normal; at other times each line was double. Pickering found that the doubling was regular, and before

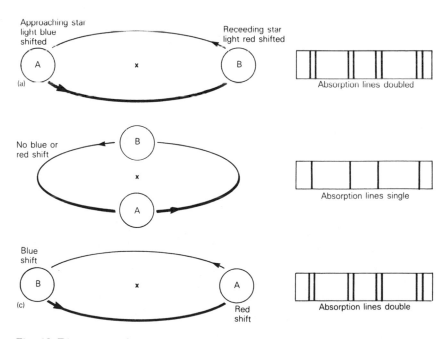

Fig. 13 Diagram to show principle of a spectroscopic binary

long he realized what was happening. Mizar A is not a single star. It is made up of two components, so close together that no telescope can show them separately.

Neglecting the overall radial velocity of the system, which can be allowed for, we have to deal with two stars moving round their common centre of gravity (Fig. 13). In the first diagram, star A is approaching us, while star B is receding; therefore star A will show a blue shift in its absorption lines, and star B will show a red shift. The lines are therefore moved to opposite sides of their mean position, and will be doubled. In the second diagram, the stars are moving 'sideways-on'; there are no shifts, and the lines are superimposed, so that they appear single. In the third diagram, star A is receding (red shift) and star B is approaching (blue shift), so that again we have a doubling effect. Pickering found that the revolution period of the Mizar A pair is 20.5386 days, and he was able to show that the average separation between the two is no more than 18 million miles, so that there is not the slightest hope of seeing them individually. But for the Doppler effect, we would not be able to tell that Mizar A is anything but a solitary star. Each component is about thirty-five times as luminous as the Sun.

Other spectroscopic binaries were soon found, by Pickering and also by other astronomers (notably by H. C. Vogel, in Germany). Today the total number runs into many thousands, and they have been of special importance to astronomers, because the ways in which they behave give clues to their real masses.

Interestingly, Mizar B – the fainter member of the visible pair – has also been found to be a spectroscopic binary. The period seems to be 182 days, so that the components are more widely separated than those of Mizar A. And to complete the picture Alcor, too, is a spectroscopic binary. This means that the Mizar system is made up of no less than six suns, and qualifies as a sort of stellar family. There is even a strong suspicion that another member of the group exists, associated with Mizar B.

What would the view be like from a planet in the Mizar system? It all depends upon the planet's position; the sky might well be illuminated by several suns at the same time, so that there would be fascinating shadow effects. Of course, we do not know whether any such planet exists, and the evidence on the whole appears to be rather against it; but we cannot be sure.

Mizar has many claims to fame. It is the best example of a naked-eye double; it was the first-known telescopic double; it was the first double to be photographed, and it was the first star found to be a spectroscopic binary. There is also the still intriguing minor mystery of why the Arabs described it as a test of eyesight. I suggest that you go outdoors on the next dark, clear night, find the Plough, identify

Mizar, and then look for Alcor. I will be very surprised if you fail to see it at the first glance. And if you have even a small telescope, you will be able to see the entire group: the two bright Mizars, Alcor, and even Ludwig's star between them, masquerading as a true member of the Mizar family.

IV BETELGEUX: THE LIVES OF THE STARS

If one had to choose a candidate for the title of 'the most splendid constellation in the sky', my personal vote would go to Orion, the Hunter. Not everyone will agree. Perhaps the Great Bear is the more distinctive; the Milky Way is richer in Sagittarius, the Archer; there is much to be said in favour of the Scorpion, whose long line of bright stars really does conjure up the impression of the insect after which it is named. In the far south we have not only Centaurus, the magnificent Centaur, but also the Southern Cross, which may be more like a kite than a cross, but which cannot be misidentified. Yet to me, Orion's claims are paramount. It contains two of the brightest stars in the sky, Betelgeux and Rigel; its pattern, with the three stars of the Belt and the mistiness of the Sword, is unmistakable; and it is visible from all over the world, because the celestial equator passes very close to the northern star of the Belt.

It is not always on view. During parts of northern summers and southern winters it is too near the Sun in the sky to be seen at all, but around September it starts to reappear before dawn, and for the months to either side of Christmas it is dominant. In mythology, Orion was a huntsman who boasted that he could kill any creature on earth – but he had overlooked the scorpion, which stung him in the heel and brought his career to an untimely end. Note that in the sky, Orion and the Scorpion are so far apart that from countries such as England they can never be above the horizon at the same time!

Orion is a useful guide to less imposing groups. Southward, the line of the Belt points to Sirius; northward, to Aldebaran in the Bull. Also in the region are Procyon in the Little Dog, the famous 'Twins' Castor and Pollux, and Capella in the Charioteer. Once Orion itself has been found, there will be little difficulty in working through his retinue.

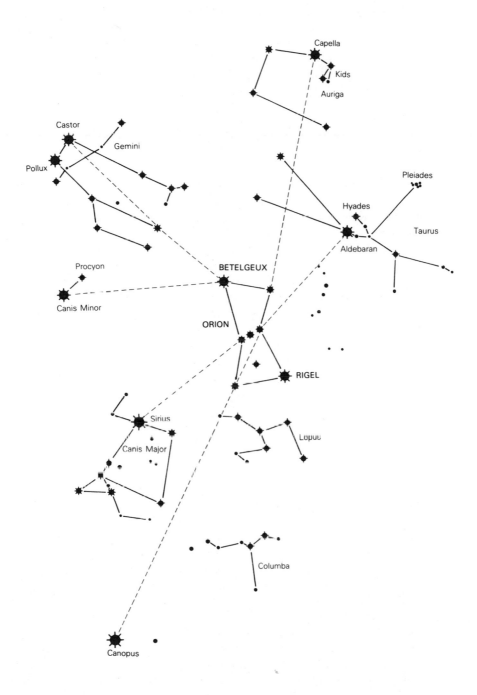

Fig. 14 Orion, showing position of Betelgeux and surrounding areas

The two leaders of Orion, Betelgeux and Rigel, are obviously different. Rigel is pure white, and has been found to be immensely luminous; you would need at least 60,000 Suns to match it. Betelgeux, by contrast, is orange-red. It is not so brilliant as Rigel, and is decidedly variable in light, but it is always of the first magnitude or brighter, and it is one of the most fascinating stars in the sky.

In theory, the brightest star of a constellation is Alpha, the second Beta; but as we have noted, there are many exceptions – and Orion is one, since it is Betelgeux which is dignified by the letter Alpha. The proper name may be spelled in several different ways, and there are also various pronunciations of it; I remember that Sir James Jeans, a great astronomer of half a century ago who was also a famous writer and broadcaster, insisted upon referring to it as 'Beetle-juice'! It is often spelled Betelgeuze or Betelgeuse; the name seems to have come from the Arabic for 'the Armpit of the Central One', originally Ibt al Jauzah, but corrupted into Bed Elgueze, Beteigeuze, and so on until we have reached the modern form. It is impossible to say which spelling is correct, if only because the Arabs did not use the Roman alphabet, and one has to depend upon phonetics. I have used the final -x; if you prefer -se or -ze, it comes to the same thing.

There is no doubt that Betelgeux changes in brightness. It is in fact the most brilliant of all the accepted variable stars (unless we count the unique Eta Carinæ, about which I will have more to say in Chapter 12, and which has not been visible with the naked eye for a hundred years now). At its peak, Betelgeux is nearer magnitude 0 than 1, and is comparable with Rigel and Capella (each 0.1). When faintest, it is not very different from Aldebaran (0.8) or even Pollux (1.1). The *Cambridge Sky Catalogue* gives the extreme range as 0.4 to 1.3, but it is certainly true that Sir John Herschel, son of the discoverer of Uranus, noticed the light changes as long ago as 1836, and commented that on a few occasions he had seen it shine as 'the largest star in the northern hemisphere', which would put it at around 0.0. I have been following its fluctuations ever since 1936, and I have seen it rise to near-equality with Rigel, though I have never seen it sink to the dimness of Aldebaran.

Of course, naked-eye estimates such as mine are bound to be comparatively rough. The method is to compare the variable with other stars which do not change, and which can be used as comparisons. This is quite satisfactory when the variable and the comparison star are close together, as with a telescopic variable, when the field of view is no more than a degree or two; but when you try to compare Betelgeux with, say, Capella you have to take the altitude difference into account. The lower a star is, the more of its light is absorbed by the Earth's atmosphere, and the fainter it will look (Fig. 15). The effect is termed 'extinction', and can be calculated, but it is hard to allow for when the variable and the comparison star are wide apart. Suppose, for instance, that I go out on a winter evening in England and look first at Betelgeux and then at Capella, finding them equal? Capella is then almost overhead, Betelgeux much lower down, so that if they appeared to be the same it would follow that Betelgeux would actually be the brighter. It is also decidedly tricky to compare a reddish star with a white one, which is why the orange Aldebaran is so useful in trying to follow the fluctuations of Betelgeux. Unfortunately Betelgeux is nearly always much the more brilliant of the two, which makes estimates even more difficult.

Betelgeux is classed as a semi-regular variable, and the *Sky Catalogue* gives its period – that is to say, the interval between successive maxima – as 2110 days, or 5.8 years, but this period is at best very rough, and is subject to marked irregularities. In fact one can never tell just how Betelgeux is going to behave, and this brings us on to the question of why it fluctuates at all. Presumably it must be unstable; but is it a very young star, still settling down after being formed, or is it very advanced in its life-story, so that it is suffering from the effects of old age?

When we set out to explain how the stars are born and how they

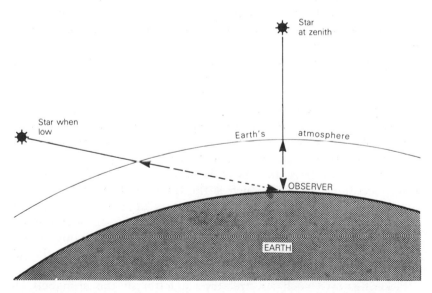

A star appears fainter when low because it is viewed through a greater thickness of atmosphere. It also appears to twinkle more strongly.

Fig. 15 Principle of extinction

evolve, we are faced with the problem that in the universe most things happen very slowly. There are exceptions, of course, such as in the violent outbursts which we call novæ and supernovæ, but in general the time-scale is long. Normally it is quite hopeless to expect to see a star change on a permanent basis. Go back to the days of the Trojan War, and the stars would be virtually the same as they are now. Go back a few million years, and there really would be changes; but since we have no access to a time machine, we must do our best by observing the stars as we see them today.

I have worked out a comparison which I have given before, but which I repeat here because I have never been able to think of anything better. When we try to tell which stars are young and which are old, our only method is to select stars which are obviously in different stages of their careers and then put them in order. Picture an intelligent being from another solar system who makes a brief visit to Earth, and spends a couple of hours in, say, Oxford Street or Times Square before returning to his (or hers, or its) own world. He will see babies, boys, youths, men and old men; if he is observant, he will be able to realize that a baby turns into a boy, and so on. In the end he will be able to work out the evolutionary cycle of a male human being – though unless someone has been kind enough to tell him about the facts of life, he will have no idea of how the baby can have been produced in the first place.

Our hypothetical visitor would be in a rather better situation than we are, because it would be fairly plain that a baby is younger than a man. But how do we go about this with regard to the stars? Which are infants, and which are senile?

The essential key is colour, and the colour of a star depends upon its surface temperature. By examining the spectra, we can find out the temperatures. That of the white Rigel is over 11,000°C; our yellow Sun, around 6000°; the orange-red Betelgeux, a mere 3000° or so. The luminosity, of course, depends upon size as well as temperature, and we now know that Betelgeux is very large indeed.

I have already said something about the spectra of the stars. With a normal star we have a rainbow background, upon which are dark absorption lines, each of which is the trade-mark of one particular substance. At a very early stage in the story of stellar spectroscopy, attempts were made to divide up the stars into various well-defined spectral types. Preliminary efforts were unsatisfactory, but in 1890 Edward Pickering produced a scheme which is still in use, though admittedly it has been modified and amended several times.

Pickering gave each spectral type a letter of the alphabet. He meant to start with the hottest stars (letter A) and work through to the coolest, but the final result was alphabetically chaotic, because some of the original types, such as E, proved to be unnecessary.

Today we accept that most of the stars are included in six classes:

B: hot bluish-white stars; surface temperatures up to 80,000°C*
A: white stars, up to 25,000°
F: yellowish-white, up to 7500°
G: yellow, up to 6000°
K: orange, up to 5000°
M: orange-red, up to 3400°

Typical stars are Spica in Virgo (B), Sirius (A), Polaris (F), the Sun and Capella (G), Arcturus in Boötes (K) and Betelgeux (M). There are a few other classes – W and O, hotter than B, and R, N and S, cool and reddish – but for the moment they need not concern us.

There is no difficulty in telling which type of spectrum is which. For example, lines due to the element helium are dominant in B-type stars; with A stars, it is hydrogen which shows up best. When we come to the comparatively cool stars, we have to deal not only with lines due to atoms, but also to those produced by atom-groups or molecules. Molecular lines do not occur in the spectra of 'early-type' stars (B and A) because the temperatures are too high; molecules would be broken up at once.

All this had become apparent by the start of our own century. By then we also knew the approximate luminosities of many of the stars. If we could find out the distance of a star, then its luminosity followed; if not, then it was often possible to work out the power of a star by studying the characteristics of its spectrum. Then, shortly before the First World War, the Danish astronomer Ejnar Hertzsprung made a very interesting discovery. He drew a diagram, linking the temperatures of the stars with their real luminosities, and came to an unexpected result. Because similar work was being carried out at the same time by Henry Norris Russell, in America, the diagrams are known as Hertzsprung-Russell or HR diagrams. Their importance in modern astronomy can hardly be overestimated.

In the typical HR diagram, shown in Fig. 16, we see luminosity plotted against spectral type, which comes to the same thing as plotting it against surface temperature; remember, B stars are the hottest and M the coolest (neglecting the rare types at opposite ends of the chief series). There is a well-defined 'Main Sequence' from the upper left to the lower right; at the upper right there are very luminous stars with cool surfaces, known as giants and supergiants, while at the lower left there are the hot but feeble white dwarfs. The pattern is so striking that it cannot possibly be dismissed as

---

*From here on I have given all temperatures in degrees Centigrade (or, if you prefer it, degrees Celsius).

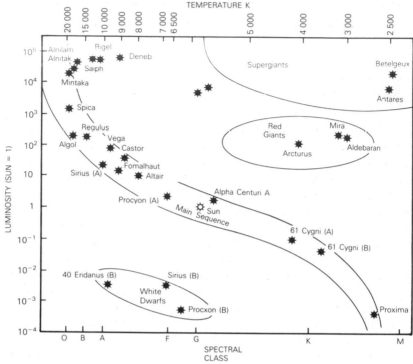

Fig. 16   HR diagram

coincidence. There must be a definite reason for it, and this brings us on to the ways in which the stars shine.

All kinds of theories were proposed at various times, most of which sound decidedly peculiar today. For example, it was suggested that the Sun was constantly 'fed' by pieces of material falling into it from space. Rather better was the idea that it was simply burning, in the manner of a vast furnace, but this had to be given up when it became known that the Earth – and hence the Sun – is very old. The age of the Earth can be estimated in several different ways, all of which lead to the same result: around 4600 million years. This is indeed a long time. Represent the age of the Earth by one calendar year, and you will have to compress the entire story of mankind into the last hour of December 31, which is indeed a staggering thought.*

*The age of the Earth caused considerable discussion over the years. From Biblical studies, the seventeenth-century Archbishop of Armagh, Henry Ussher, calculated the date of creation as being 23 October BC 4004, and even gave the exact time. The revelation that fossils were millions of years old did not perturb him or his followers in the least, and the BC 4004 date was still being supported by Dr John Lightfoot, Vice-Chancellor of Cambridge University, as recently as 1859. Luckily, views at Cambridge have changed somewhat since then!

It is easy to calculate that a Sun made up of coal, burning fiercely enough to send out as much heat as the real Sun actually does, would not last for long on the cosmic scale. It was next suggested that the Sun and other stars radiate energy because they are slowly shrinking. It is true that gravitational energy would be released in this way, but again the time-scale is all wrong, and is very much too short. Russell himself put forward the idea that the energy source of the stars could be due to the annihilation of matter, so that certain types of particles were wiping each other out and releasing energy in the process. This was a decided improvement, but led to a life-cycle of millions of millions of years, which was as obviously too long as the previous results had been too brief.

All the same, Russell's theory did seem to fit in well with the HR diagram, and this is where we come back to Betelgeux.

It was (and still is) assumed that a star begins its career inside a cloud of gas and dust in space known as a nebula. At first it is cold, and is merely part of a huge, incredibly rarefied mass. But any chance condensation will lead to contraction; gravity takes over, and as the centre of the mass becomes denser it also heats up. Eventually it reaches a temperature high enough to make it shine. Our cold cloud has become a true star.

At first, said Russell, the star would be very large and very cool by stellar standards; it would also be unstable, because it would not have settled down, and it would fluctuate in light. We would be dealing with a red giant such as Betelgeux, preparing to embark upon the most glorious part of its career. As it shrank, it would go on heating up; it would become an orange giant, then a yellow giant, and would then join the Main Sequence at the upper left of the HR diagram. It would now be much smaller, but with a very hot surface.

Next it would continue to shrink, but would also start to cool down. It would slide along the Main Sequence, passing through the orange and yellow stages before turning into a small, cool red dwarf. Its final fate would be as a cold, dead globe from which all light and heat had departed. Of course the process would not be rapid, but eventual extinction would be certain.

Note that in the HR diagram, the red and orange stars, and to a lesser extent the yellow stars also, are divided sharply into two groups: giants and dwarfs. We have red giants such as Betelgeux, thousands of times more powerful than the Sun, and red dwarfs such as Proxima Centauri, the nearest star beyond the Solar System, which have only a tiny fraction of the Sun's luminosity. Red stars of about the same power as the Sun do not seem to exist at all. Everything looked very neat and tidy – particularly as the white dwarfs were not then known, and in any case come into a completely different category.

According to Russell's theory, therefore, Betelgeux was a very young star. It was unquestionably huge; the diameter is now known to be of the order of 250 million miles, which is big enough to swallow up the entire path of the Earth round the Sun. The distance is given as 520 light-years, and the peak luminosity is about 15,000 times that of the Sun. Certainly this is only a quarter of the output of Rigel, but the vastness of Betelgeux is balanced to some extent by its relative coolness.

Astronomers became confident that they were working along the right lines, and that they were now able to arrange the stars in order of age – or, rather, in order of their stage of evolution. Red giants such as Betelgeux were young; yellow giants such as Capella rather older; Rigel and Sirius older still; the Sun well into middle age, and Proxima well on the way to extinction. It was only gradually that doubts started to creep in, and not for a further two decades were astronomers forced to realize that they had picked wrong. Betelgeux was not young at all. It was, by stellar standards, ancient.

The crux of the whole matter was the source of stellar energy, and as astronomers learned more about the make-up of matter they had to admit that the annihilation process described by Russell simply would not work. In 1939 two theorists proposed new ideas at almost the same time, Hans Bethe in America and George Gamow in Europe. Bethe's discovery was made in a rather curious way. Apparently he had been attending a scientific meeting in Washington, and was returning to Cornell University by train when he began thinking about the reactions taking place inside stars. By the time lunch was served, he had solved the main problem. That must have been one of the most significant train journeys in the history of astronomy!

What Bethe (and Gamow) realized was that the source of a star's energy is not annihilation, but what we now call nuclear transformation. An atom is made up of a central nucleus, around which move less massive particles termed electrons. (This, I admit, is a gross over-simplification; in particular, it is impossible to regard a nucleus or an electron as a solid lump, but the general picture is good enough for the moment.) Hydrogen, the lightest of the elements, has an atom made up of a nucleus with one circling electron; helium, the next lightest element, has two electrons, and so on. There is a complete sequence from hydrogen up to the heaviest element known to occur in nature, uranium, whose complicated nucleus is surrounded by no less than 92 electrons.

In a star such as the Sun, there is more hydrogen than anything else; in fact, in the universe as a whole, hydrogen atoms outnumber those of all the other elements combined. But deep inside the Sun, where the temperature is tremendous – at least 14 million degrees, possibly more – the atoms are broken up, and we are left with 'stripped' nuclei and unattached electrons. The hydrogen nuclei are running together to make up nuclei of the next element, helium. It takes four hydrogen nuclei to make one nucleus of helium, and each time this happens a little energy is released and a little mass is lost. The mass-loss amounts to 4 million tons per second, but even so there is so much material in the Sun that there is no danger of its changing much for a very long period in the future.

Now we can take a new look at the whole picture. As before, we assume that a star begins by condensing out of a rarefied interstellar cloud, and will at first be cool, unstable and reddish, but it will not be a powerful giant such as Betelgeux. Its later career will depend mainly upon its original mass. If this mass is less than about a tenth of the Sun's, the star will never become hot enough for nuclear reactions to start inside it (around 10 million degrees is needed), and it will simply glimmer feebly for a long period before fading away. Nowadays we have identified 'failed stars' of this kind; they are known, rather misleadingly, as brown dwarfs (Fig. 17).

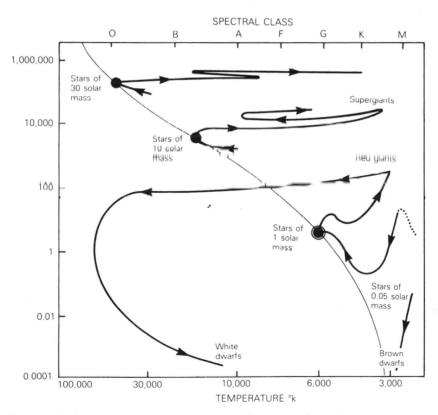

Fig. 17 Evolutionary tracks of stars, according to modern theory

With a star of solar mass, the core temperature will rise until the hydrogen-into-helium process is triggered off, and the star joins the Main Sequence at a point which depends upon its mass. Thus the Sun has joined the Main Sequence at type G; Sirius, considerably more massive, has joined it at type A, to the upper left of the HR diagram. Contrary to Russell's view, a star does not slide up or down the Main Sequence. It simply stays more or less where it is in the HR diagram until it starts to run short of its essential fuel, hydrogen, when it will have to change its structure drastically.

When this happens, the star will have to start using different fuels, and building up heavier and heavier elements inside it. It will leave the Main Sequence, and move into the giant branch of the HR diagram. To go into full detail would take many pages, and for the moment it is enough to say that after all the nuclear reserves are exhausted, a solar-type star collapses to become a white dwarf, small in size but incredibly dense. I will have more to say about white dwarfs in the next chapter.

Betelgeux, much more massive than the Sun, is running through its life-cycle much more quickly. In absolute terms it is not as old as the Sun, but it has evolved relatively rapidly. In, say, 4000 million years hence the Sun will still be on the Main Sequence, but disaster is bound to have overtaken Betelgeux.

With Betelgeux, the original condensation out of the interstellar cloud proceeded at accelerated pace; then came the Main Sequence stage, with hydrogen being used up, and then came a period when heavier elements were being used as fuels – something which cannot happen to the Sun in the same way, even in the far future, because the core temperature will never rise high enough. At present Betelgeux has a structure quite different from that of a Main Sequence star. There is a core at a temperature of perhaps 3000 million degrees, surrounded by 'shells' of various types and then a huge, unstable 'atmosphere', which is what we actually see. Betelgeux is swelling and shrinking, changing its output of energy as it does so.

Also, Betelgeux is the closest to us of the red supergiant stars, a supergiant being an exaggerated version of a giant. It is surrounded by a tenuous shell of material with an apparent diameter of 3 minutes of arc, expanding at well over 2000 miles per hour, so that the material at its outer edges must have left the star a mere 10,000 years ago. There is some evidence that the shell is not symmetrical, in which case its shape must be affected by the movement of Betelgeux through the very tenuous interstellar medium. It has also been estimated that at the moment Betelgeux is losing mass at a rate equivalent to the whole mass of the Sun in no more than 100,000 years. Even a supergiant cannot tolerate such a huge loss for a protracted period, which is why Betelgeux, in its present form, cannot last for nearly as long as our relatively mild Sun.

Because Betelgeux is so large and is the closest of the supergiants, efforts have been made to take photographs showing surface detail on it. This is a very difficult problem, because even Betelgeux has an apparent diameter of less than 0.06 of a second of arc, but use has been made of a comparatively new technique known as speckle interferometry, which involves combining a great many short-exposure photographs and then analysing them electronically. If the exposure is below a twentieth of a second we can more or less ignore the smearing effects of the Earth's air, and the technique really works. There have been various published pictures which, it is claimed, show that Betelgeux has dark patches on its surface, though as yet it is too early to say whether these pictures are valid or not.

The volume of Betelgeux is at least 160 million times that of the Sun, but the mass is only about 20 times that of the Sun, and the mean density of the globe is less than 1/10,000 of that of the Earth's air at sea-level, corresponding to what we normally call a laboratory vacuum. So what would the star look like from close range? Its outer limb would not be sharp and clear-cut, as with our Sun, because the gases there are so much more rarefied; the overall impression would presumably be that of an orange-red mass, with darker regions here and there. However, it is very doubtful whether there could be any planets in the Betelgeux system. If there had originally been any, they would by now have been destroyed, because Betelgeux is much more luminous as a supergiant than it must have been when it was a Main Sequence star.

Finally, what will be the eventual fate of Betelgeux? Long before our Sun dies, and long before Man vanishes from the Earth, Betelgeux in its guise of a red supergiant will have left the cosmic stage. It may well explode as a supernova. But for the moment it is there for our inspection, and for several months in every year you can go outdoors at night and gaze at the lovely, orange-red star in the shoulder of the Hunter.

V SIRIUS: THE DOG-STAR AND THE PUP

Sirius, the leader of Canis Major – Orion's senior Dog – is much the most brilliant star in the sky. It is three-quarters of a magnitude brighter than its nearest rival, Canopus in the far south, and it is almost a magnitude and a half brighter than Arcturus, which outshines any star in the northern hemisphere of the sky. True, Sirius owes its pre-eminence to the fact that at a distance of only 8.6 light-years it is one of our nearest stellar neighbours, but it is none the less glorious for that.

The name, surprisingly, is not Arabic, like most of the other star

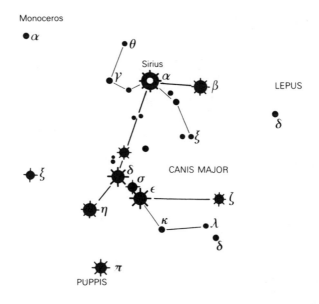

Fig. 18 Position of Sirius

names. It comes from the Greek for 'sparkling' or 'scorching', and the correct pronunciation is Sī-rius with a long first *i*, though most people prefer to call it 'Sirri-us' (and after all, what does it matter, except to pronunciation fanatics?). Because it is in the Great Dog, it is often known as the Dog-Star, and it has been mentioned many times throughout history. In Ancient Egypt it was called Sothis, and was regarded as most important. At the time of its heliacal rising – that is to say, the time in the year when it could first be seen in the dawn sky – the Egyptians knew when to expect the annual flooding of the Nile, which was so vital to their whole economy.

Sirius is a white star, but when low down it gives the impression of flashing various colours of the rainbow. The phenomenon of twinkling, or scintillation, has nothing directly to do with the stars themselves, and is caused solely by the Earth's unsteady atmosphere (Fig. 15). Sirius shows it particularly strongly, because it is so brilliant and because (like all stars) it is virtually a point source of light.

One mystery about Sirius is that many of the astronomers of ancient times, including Ptolemy, stated categorically that it was red. It is certainly not red now, and I will return to this question later. Meanwhile, what about the power of Sirius, and what kind of a star is it?

Its spectrum is of type A, and it lies on the Main Sequence. It has one of the largest proper motions known, and Edmond Halley, in 1718, found that it was one of three brilliant stars to have changed position perceptibly in recorded times (the others were Arcturus and Aldebaran, both of which, however, are much farther away than Sirius). The real distance of Sirius works out at around 50 million million miles (about 80 million million km), so that of all the first-magnitude stars only Alpha Centauri is closer. It is at present approaching us at the rate of 4½ miles (8 km) per second.

The diameter of Sirius is not far short of twice that of the Sun: around 1.7 million miles. Its luminosity is equal to that of twenty-six Suns; its surface temperature is of the order of 10,000°C, and the temperature at the core may be as much as 20,000,000°C, appreciably higher than that of the Sun. Sirius is also about two and a half times more massive than the Sun, and therefore it is squandering its fuel reserves more quickly, though it still has a long time to go before it leaves the Main Sequence and moves into the giant branch of the HR diagram. Like all Main Sequence stars, it is officially classed as a dwarf. If it were as far away as (say) Rigel, a powerful telescope would be needed to show it at all – while if Rigel were as close to us as Sirius, it would cast strong shadows.

Unlike the Sun, Sirius is not a single star. It has a companion, which was actually predicted well before it was actually seen.

I have already talked about the career of Friedrich Wilhelm Bessel, who was the first man to measure the distance of a star, and who might well have been the first to predict the position of the hitherto unknown planet Neptune had not illness struck him down when he was at the height of his powers. Bessel, using the Königsberg heliometer, made a series of careful measurements of the position of Sirius in the years following 1834. There was no problem in following the proper motion, which amounts to over 1.3 seconds of arc per year, so that since the time of Ptolemy it has travelled against its background by an amount equal to about one and a half times the apparent diameter of the full moon. But Bessel found something very significant. Sirius was not moving in a regular way. It was 'weaving' its way along, as though it were being perturbed by a companion body, which Bessel could not find (Fig. 19). Could it be a planet? No: to have a measurable effect upon a star such as Sirius, the invisible object would itself have to have a mass comparable with that of the Sun, which is much too heavyweight for a planet. So presumably the companion had to be a star; but despite all his efforts Bessel could never catch a glimpse of it. All he could say was that it probably existed, and had an orbital period, with respect to Sirius itself, of about 50 years. After Bessel's death another talented mathematician, C. H. F. Peters, made similar calculations and came to much the same result, but still the companion remained elusive.

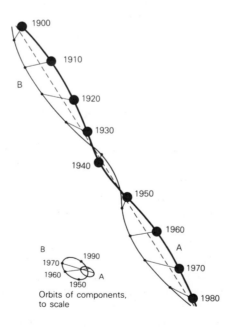

Fig. 19 Proper motion of Sirius

Then, in January 1862, one of America's most experienced telescope-makers, Alvan G. Clark, completed a new refractor with an 18½-inch object-glass; it was then the largest refractor in the world, and Clark was pleased with it. (It is still in full use, at the Dearborn Observatory in Illinois; I have observed with it myself.) One of Clark's test objects was Sirius, whose great brilliance will show up any tiny imperfections in the optics and produce the dreaded telescopic 'ghosts'. There was indeed a speck of light near Sirius, but it was no ghost; it was the long-expected companion, almost exactly in the position which Bessel and Peters had predicted. Because Sirius has so often been called the Dog-Star, the companion, officially Sirius B, has been nicknamed the Pup!

It is not particularly faint. Its magnitude is 8.6, so that if it were isolated it would be an easy object in good binoculars. What makes it so difficult is that it is completely drowned by the glare of its primary, which outshines it by a factor of 10,000. The angular separation from Sirius A ranges between 3 seconds of arc and as much as 11.5 seconds of arc. The separation was at its greatest in 1975 (see Fig. 20), and for some years around that date I had no real difficulty in seeing the Pup with my 15-inch reflector, provided that conditions were good, with Sirius at its maximum altitude above the horizon, and that I hid the glare of the bright star by using an 'occulting bar' in the optical system of the telescope. Now, in the mid-1980s, it is becoming more difficult. The next closest approach will be in 1994, and the orbit has turned out to be quite eccentric, more like that of a comet than that of a planet.

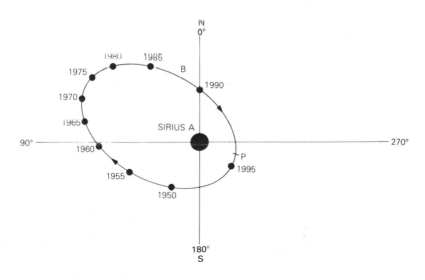

Fig. 20 Orbit of Sirius B, with dates

Yet the relative movements of the Dog and the Pup seemed at once to indicate that they were not so very unequal in mass. Of course the Dog was the 'heavier', but it is a characteristic of all stars that the smallest bodies are the densest; as we have seen, Betelgeux in Orion has a diameter more than 250 times that of the Sun, but only about twenty times the solar mass. Presumably, then, the Pup would be large, cool and red. This was certainly the view of a famous Irish astronomer, J. Ellard Gore, who also wrote some popular books which were unusual in content and make fascinating reading even today. In 1907 he published his *Astronomical Essays*, in which he gave perfectly logical reasons why the Pup must be at a low surface temperature:

> If its faintness were merely due to its small size, its surface
> luminosity being equal to that of our Sun, the Sun's diameter
> would be the square root of 1000, or 31½ times the diameter of
> the faint star, in order to produce the observed difference in light.
> But on this hypothesis the Sun would have a volume 31,500 times
> the volume of the star, and, as the density of a body is inversely
> proportional to its volume, we should have the density of the
> Sirian satellite over 44,000 times that of water (the Sun's density
> being 1.5). This, of course, is entirely out of the question, and the
> result shows at once that the luminosity of the satellite's surface
> cannot possibly be comparable with that of the Sun. Its surface
> must be enormously less luminous than the Sun's surface.

Gore added: 'The satellite has cooled down considerably, and is probably far advanced in regard to the total extinction of its light. It is unfortunate that its spectrum cannot be observed, as it should be a most interesting one.'

In this latter statement Gore was right – though, sadly, he did not live to see it (he was killed in a street accident in Dublin in 1910; he was only 55 years old). In 1915 W. S. Adams, at Mount Wilson in California, managed to obtain a spectrum of the Pup. The result came as a true astronomical sensation. The surface of the star was not cool at all; it was hot – much hotter than the Sun's, and corresponding to a type of late F or early A.

If the Pup were hot but faint, then there was only one conclusion: it must be small. The diameter works out at about 26,000 miles (about 40,000 km), which is not much more than three times that of the Earth and is smaller than that of a planet such as Uranus or Neptune. Yet the movements relative to the primary prove that the mass is about the same as that of the Sun (the latest estimate is 98 per cent). This means that a vast amount of mass, over 300,000 times that of the Earth, has to be packed inside a planet-sized globe, and the density rockets to an unbelievable level. Gore's

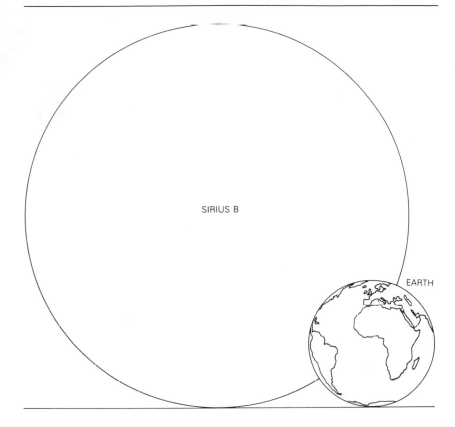

Fig. 21 Size of Sirius B, compared with Earth

value of 44,000 times that of water has proved to be not 'out of the question', but a gross understatement. The real value is more like 125,000 times that of water, and the luminosity is a mere 1/400 of that of the Sun. Sirius B, the Pup-star, is what is termed a white dwarf.

We have come across white dwarfs before; they lie at the lower left in the HR diagram, but are not conventional dwarfs in the sense that the feeble red and yellow stars are. A white dwarf is a bankrupt star, which has used up its reserves of nuclear fuel. To explain this, we must go back to the story of stellar evolution.

With a star of mass similar to the Sun, we begin with condensation from an interstellar cloud or nebula; we have shrinking, heating-up, the triggering-off of nuclear reactions at a core temperature of 10,000,000°C, and then a long period of steady, stable existence on the Main Sequence. When the supply of hydrogen fuel runs low, different reactions occur: the star expands, with a shrinking, hotting-up core and a cooling surface, to become a red giant (not a red supergiant such as Betelgeux; this involves a star

51

of more than about 1.4 times the mass of the Sun, for reasons which need not concern us at the moment). Next, the star throws off its outer layers entirely to produce what is called a planetary nebula. The term is a bad one, because a planetary nebula is not truly a nebula and is certainly not a planet: it consists of the core of the old star and a discarded shell of gas which gradually expands into space until it ceases to be visible. Many planetary nebulæ are known, and a few, such as the famous Ring Nebula in Lyra, are within the range of small telescopes.

The core of the old star has 'shrunk in' on itself. Ordinarily, the atoms which make up all material are composed of particles together with empty space, rather like a miniature Solar System, but in the core of a white dwarf the atoms are broken up, and all their constituent pieces are crammed together with almost no waste of space, which accounts for the amazingly high densities.

In trying to explain this sort of behaviour, there is one problem which I have never been able to overcome: that of giving an intelligible account without falling into the trap of assuming that the parts of atoms (nuclei and electrons) are solid lumps. This is not the case, and if you are content to bear it in mind I can at least give an analogy which may not be too misleading. Picture the balls on a snooker-table, set out at the start of a frame of snooker; they cover much of the table, with wide spaces between them. Now collect all the balls and cram them together so that they are touching each other. They take up much less room, but they are the same snooker balls as before. In a white dwarf star, the core of a star which was once of the same type as the Sun, this is the sort of thing that has happened, producing what is termed 'degenerate matter'.

Because white dwarfs are still hot, they continue to shine feebly. They will continue to do so for a vast span of time, and it will be many thousands of millions of years before they become cold and dead. Completely dead stars have been called black dwarfs, but we are by no means sure that the universe is yet old enough for any of them to have been produced, and obviously we could not hope to see them, because they would sent out no radiation at all. However, there are many white dwarfs; *en passant*, Procyon, the leader of the Little Dog, also has a white dwarf companion which was predicted by Bessel. We know of some white dwarfs which are much smaller and denser than the Pup, and a few which are even smaller than the Moon. There seems no escape from the conclusion that all normal Main Sequence stars will become white dwarfs eventually, and this includes the Sun, though by then human beings will not be there to watch; our world can hardly hope to survive the blast of radiation which the Sun will send out during its red giant stage.

There are two mysteries associated with Sirius, one of which is worth taking seriously while, to be candid, the other is not. Taking the latter first, we may cast a brief look at the Dogon tribe in the state of Mali in Africa. I have never been to Mali,* and neither has Robert K. Temple, a writer who published a book in 1975 claiming that the Dogon knew all about the Companion of Sirius, even to a good estimate of its 50-year period. In the book it was also said that there were two French scientists who had recorded the Dogon theory that Sirius B consisted of 'sagala', a material much heavier than any metal on Earth. The Dogon were also credited with knowledge of Jupiter's satellites and the ring system of Saturn.

Any theory, no matter how weird, is bound to gain its supporters; one has only to think of the flying saucer craze, which began in the 1940s, was frenetic in the 1950s and is not completely dead even yet. But since in Dogon lore everything has its 'double', Sirius, the brightest star in the sky, would not be expected to be an exception; and since French missionaries had been in Mali for some time, no doubt meddling in Dogon affairs, it is quite on the cards that something about astronomy would have been said.

The other mystery, which is very different, concerns the colour of Sirius. Today it is pure white. Yet as we have already noted, many astronomers of ancient times called it red. To the Greeks and Romans it was an unlucky star; in Virgil's Æneid we read of 'the Dog Star, that burning constellation, when he brings drought and diseases on sickly mortals, rises and saddens the sky with inauspicious light'. Moreover, the name, which means 'scorching', would not normally be associated with a white star.

The last great astronomer of Classical times was Ptolemy, who was active around the year AD 150, and who lived in Alexandria, which had by then succeeded Athens as the cultural centre of the civilized world. Ptolemy's major work, which has come down to us by way of its Arab translation (the *Almagest*) is really a summary of all the scientific knowledge of the time, and it contains a star catalogue. Periodical attempts to discredit Ptolemy have been signally unsuccessful, and he was undoubtedly an excellent observer. He listed six bright stars which were red: Arcturus, Aldebaran, Pollux, Antares, Betelgeux, and – Sirius. Of these, Pollux is obviously yellowish or 'off-white'; Arcturus is light orange, Aldebaran and Betelgeux orange-red, and Antares very red. This leaves Sirius as the odd one out.

If this were all, we would simply have to admit that on this occasion Ptolemy either made a mistake or else has been mistrans-

---

*I once considered going to Mali to observe a total eclipse of the Sun, but I was warned that although I might be able to get in it would certainly be much more difficult to get out again. Prudently, I went elsewhere.

lated. But Ptolemy was not alone. Homer, in the *Iliad*, seems to compare Sirius's light with the gleam on the copper shield carried by Achilles; the poet Horace, around 65 BC, wrote that 'the red Dog Star divides its children', and there are significant comments also made by Pliny and Ovid. Seneca, who lived from 4 BC to AD 65, wrote that 'the redness of the Dog Star is more burning; that of Mars is milder; Jupiter is not red' – and if Sirius were redder than Mars, the colour would have had to be remarkably pronounced.

All this is very interesting. Was Sirius really red? It was white by the tenth century AD, as we know from the description given by the great Arab observer Al-Sufi. We are dealing with records dating back some two thousand years.

Apparently the first near-modern writer to suggest that Sirius was once a red star was Thomas Barker, in 1760. Barker qualifies as a scientist, and wrote voluminously about subjects ranging from mathematics to theology and vegetarianism; not much of his work is remembered today, but he did collect all the old references to Sirius, and became convinced that there had been a colour-change. He was later supported by Baron Alexander von Humboldt, who wrote in his *Cosmos* (1845) that Sirius was 'the one example of a star historically proved to have changed colour'. Next in the story came Dr Thomas John Jefferson See (always known as, T. J. J. See), who took up the case energetically, first in 1892 and then in 1927.

T. J. J. See, born in 1866, must have been a curious person. He came from Missouri, and worked successively at the observatories in Berlin, Yerkes and Lowell before becoming Professor of Mathematics in the US Navy in 1899, subsequently being appointed director of the Naval Observatory. He carried out a great deal of work with respect to double stars and stellar evolution, and also observed the planets, claiming to have seen craters on the surface of Mercury long before they were recorded by the Mariner 10 spacecraft in 1973. But to pretend that See was popular with his fellow astronomers would be what Winston Churchill once called a 'terminological inexactitude'. In fact, it may well be that there has never been an astronomer who aroused such universal dislike. In 1898 he was firmly requested to resign from the Lowell Observatory at Flagstaff, in Arizona, and the general opinion of him was aptly summed up by one of his erstwhile colleagues, A. E. Douglass: 'I have never had such aversion to a man or beast or reptile or anything disgusting as I have had to him. The moment he leaves town will be one of vast and intense relief, and I never want to see him again. If he comes back, I will have him kicked out of town.'

It is certainly true that See was eccentric. He made a habit of claiming, as his own, discoveries which had been made by other people; his sense of self-importance was as great as with any modern politician, and he even hinted that he regarded himself as immortal, so that when he finally died, at an advanced age, he must have had a tremendous shock.* But in spite of all his personal quirks, there are reasons to believe that See was a capable observer – we cannot even discount his Mercurian craters, though personally I admit to grave doubts about them – and he was also a capable historical researcher, though he did his reputation a great deal of extra damage in his later years by his strong opposition to Einstein's theory of relativity.

Be this as it may, See painstakingly collected twenty reports of the alleged redness of Sirius, made between the periods of Homer and Ptolemy. The situation is worth looking at, and we cannot merely dismiss it out of hand, as we can the Dogon theory.

If we are prepared to believe in the old reports, then there can be only a few possible explanations:

1 Sirius A really has changed from a red star into a white one.
2 It was Sirius B, the Pup, which used to be red, and which then outshone Sirius A.
3 Some cloud in interstellar or interplanetary space used to lie between Sirius and ourselves, thereby making Sirius look red.
4 Sirius was not really red, but gave that impression because of its strong twinkling in various colours.

Sirius is a perfectly stable Main Sequence star. It is not of the type to show rapid changes, and a change from red to white over the past two thousand years would be very rapid indeed by cosmic standards. If it had happened, then we would have to re-cast the whole of modern astrophysics, and for this there does not seem to be the slightest justification, so that on all counts we can at once reject explanation (1).

When we come to consider the Pup, we must be more careful. Before a star becomes a white dwarf, it must go through the red giant stage, and there is every reason to believe that the Pup did so. In that case it would have shone very brightly, and it would have been very red. At first this seems to give a satisfactory explanation of the mystery – but closer examination shows that there are several fatal flaws in it.

The time-scale is all wrong. The change from red giant to white dwarf does not happen quickly; like most things in the universe, it is gradual by everyday standards, and would take a great deal longer than two thousand years. True, a supernova outburst

---

*In 1930 Clyde Tombaugh discovered the planet we now call Pluto. A name had to be found for it, and See suggested 'Minerva'. That is why Minerva is now named Pluto!

happens very quickly indeed, but neither candidate of Sirius is a supernova candidate.

Also, it is not hard to work out the effect of combining the present Sirius with an even more luminous red giant companion. The star would shine so brightly that it would be visible in daylight, and all the observers, including Ptolemy, would have made comments about it. There have been more recent suggestions that the Pup is surrounded by an unstable shell of hydrogen gas which can periodically but temporarily glow brightly, producing a red glare, but here too the combined brightness of the system would be unacceptably great, and in any case careful spectroscopic work has revealed no trace of such a shell. Neither does it seem possible that the Pup itself is a close binary.

It was Sir John Herschel, in 1839, who suggested the idea of a space-cloud which might pass in front of Sirius and redden its light as seen from Earth. This may be valid for the extraordinary Eta Carinæ, and it was in fact Eta Carinæ which put the idea into Herschel's mind, but it does not seem to be very plausible for Sirius, mainly because we would almost certainly still be able to detect traces of any such cloud.

We come, then, to possible effects due to the atmosphere of the Earth. Sirius does seem to flash various colours when low down, as it always is from England; but the observers quoted by Barker, See and the rest lived in more southerly latitudes, from which Sirius is higher up and twinkles less. In the early part of 1986 I was in Australia (mainly because I wanted to have the best possible views of Halley's Comet), and I remember one very dark, clear night in the Northern Territory when there were no artificial lights for many miles around, and conditions were wellnigh perfect.* Sirius passed directly overhead. I could still notice slight twinkling, but no colour.

In 1973 I decided to carry out a test of my own, which I would not claim to be at all conclusive but which may at least be of some interest. In one of my *Sky at Night* television programmes, transmitted from London, I asked viewers to look at Sirius on the next dark, clear, moonless night, when the star was at its highest, and write down the colour they believed it to be. To avoid unconscious prejudice as far as possible, I gave only very vague reasons why I wanted to know. I had over 5000 replies, and they took a great deal of analysing, but the end results were as follows:

---

*I was even able to watch the globular cluster 47 Tucanæ as it dipped below the horizon; observers will know that for this sort of observation the seeing must be exceptional. Even the lights of our car were out. To be honest, the car itself was stuck in desert sand, and we had considerable difficulty in extricating it!

Bluish or bluish-white	50 per cent
White	23 per cent
Flashing all colours	14 per cent
Greenish or greenish-white	9 per cent
Yellowish	2 per cent
Orange	2 per cent

Not one viewer saw Sirius as red, and the few who called it orange added that they had made their observations when the sky was somewhat misty.

It is most unlikely that there can have been anything unusual in the atmospheric conditions over the Mediterranean area extending from Homer's time to Ptolemy's, and in any case it would be impossible to explain why only Sirius would have been affected. It may also be significant that a Chinese record of the first century BC, made by an observer named Sima Qian, classes Sirius as white.

All in all, it is my view – and, I believe, also the view of nearly everyone else – that neither Sirius nor the Pup has changed colour in recorded times. We have to assume that the old descriptions are careless, erroneous or mistranslated. Meanwhile, the Dog Star continues to shine down in all its glory; it is a wonderful sight, even though it is no longer as important to us as it was to the ancient Egyptians who watched for its dawn rising as a sign that their life-giving river was about to flood.

VI VEGA: OTHER SOLAR SYSTEMS?

One of the loveliest stars in the sky is Vega, or Alpha Lyræ. It lies so far north that from England it is circumpolar; it can be seen from every permanently inhabited country, though admittedly from the southernmost part of New Zealand it barely rises, and it is the brightest member of my unofficial 'Summer Triangle' (Fig. 22). Its magnitude is 0.03, so that it is surpassed by only four stars: Sirius, Canopus, Alpha Centauri and (just) by Arcturus. It is steely blue in colour, and is the only really brilliant star which is genuinely worthy to be called blue rather than bluish-white. It is particularly easy to recognize, because from latitudes such as that of north Europe and the northernmost United States there are only two stars which can pass over the zenith: Vega during summer evenings, Capella during winter evenings. Capella and Vega lie on opposite sides of Polaris, and at about the same distance from it.

The name (which should be, but nowadays never is, spelled 'Wega') comes from the Arabic for 'Falling Eagle'. There is also a legend about it which comes from Ancient China. According to this story, Vega (the Weaving Girl) and Altair (the Herd Boy) were so deeply in love that they neglected their duties to the gods. Chinese deities were apparently no more charitable than those of Greece, and so Vega and Altair were condemned to be separated by the Celestial River (the Milky Way), which cannot be crossed. Certainly the Milky Way runs between the two stars, but on the seventh night of the seventh month a bridge of birds temporarily spans it, so that the lovers are allowed to have a little time together. . . .

Vega has a typical A-type spectrum.* Its distance was measured in

---

*Curiously, Vega was described as 'a very brilliant pale yellow' by the late E. J. Hartung, author of a famous book about telescopic objects. However, Hartung lived in Australia, where Vega is always very low down, and apparently never travelled north of the equator, which no doubt explains the error. There is no comparison with the age-old description of Sirius as a red star.

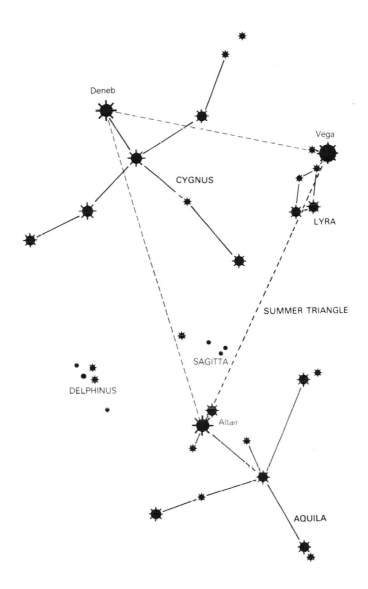

Fig. 22 The 'Summer Triangle' area

the late 1830s by F. G. W. Struve, but the result was inaccurate, because Vega is 26 light-years away – much farther than Alpha Centauri or 61 Cygni. At the moment its declination is 39 degrees north, so that it is some way from the celestial pole, but it has not always been so. The Earth is spinning rather in the manner of a gyroscope which has started to topple, and so the direction of the

59

axis describes a circle in the sky in a period of about 26,000 years (Fig. 23). At the time when the Egyptians were building their Pyramids, the north pole star was Thuban, in the constellation of the Dragon. Twelve thousand years before that, Vega was the pole star – only about 4½ degrees away from the polar point – and it will again be the pole star in 12,000 years time.

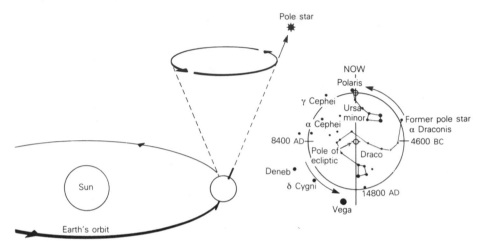

Fig. 23 Precessional movement of the north celestial pole

Vega has a surface temperature of 10,000°, and is fifty-two times more luminous than the Sun, with three times the solar mass and a very low overall density of only about one-fifth that of the Sun. It is one of the few stars near enough and large enough to have its apparent diameter measured directly, with a special instrument known as an intensity interferometer; the value is 0.0037 seconds of arc, corresponding to a real diameter of 2.8 million miles (about 4.5 million km) (as against 865,000 miles (about 1,392,000 km) for the Sun).

Vega was, incidentally, the first star to be photographed. On the night of 16–17 July 1850, a Daguerreotype of it was taken by Bond, using the 15-inch refractor at Harvard. He needed an exposure of 1 minute 40 seconds, but Vega was clearly visible.

Lovely though it looks, Vega is not particularly distinguished by its luminosity; as we have seen, it is puny indeed when compared with stars such as Deneb and Rigel. But it is of special interest because some remarkable observations were made of it in 1983 – all the more remarkable because they were so unexpected.

In January of that year IRAS, the Infra-Red Astronomical Satellite, was launched from Cape Canaveral in Florida. It was put into an orbit which took it round the Earth well above the atmosphere, and it was designed to study objects at wavelengths longer than that of light. Red light, remember, has the longest wavelength in the visible range; beyond that we come to infra-red radiations, which can be collected by means of a telescope which looks outwardly very much like a normal telescope. The main problem is that infra-red radiations from space are heavily absorbed by the Earth's atmosphere, particularly by the water vapour in it. Astronomers do their best, by taking their equipment as high as possible; Britain's largest telescope of this kind has been placed on top of Mauna Kea, the extinct Hawaiian volcano, at an altitude of about 14,000 feet. (It is worth adding that this telescope is so good that it can be used for ordinary observations as well as infra-red.) Yet this is not the complete answer, and to use space-research methods is the only real way to overcome the difficulty. Various infra-red telescopes have been launched during the past couple of decades, but IRAS was a great advance on anything previously sent up, and it proved to be remarkably successful. It functioned throughout most of 1983, and it carried out a complete survey of the sky at infra-red wavelengths, discovering thousands of new sources.

The data from IRAS were received at the Rutherford-Appleton Laboratory in Oxfordshire. Two American members of the team, Dr Hartmut ('George') Aumann of the Jet Propulsion Laboratory in California and Dr Fred Gillett of the Kitt Peak Observatory in Arizona, were using various stars as sources for calibrating the infra-red telescope aboard IRAS when Dr Gillett suddenly called out: 'Hey, Alpha Lyræ has a huge excess!' The two startled astronomers made careful checks, but other stars showed no more than the expected amount of infra-red, so that Vega really was exceptional.

The infra-red source associated with Vega was about 20 seconds of arc in diameter. Since the IRAS equipment was accurate to within 2 seconds of arc, there was no possibility of a mistake. Originally it was thought that the radiation might be coming from 'mass flowing out from Vega', but it was soon found that this would not fit the facts, because the material was not flowing away from Vega, and it had presumably been there for a very long time.

Infra-red emission comes from relatively cool objects, but Vega itself is a hot star, so that it seemed that there must be a vast low-temperature cloud of material all round the visible star. Apparently it was in the form of particles, but the particles were much larger than the tiny dust-grains found in interstellar space. Moreover, the infra-red emission came from a region extending out to about 7.5 million miles (12 million km) from Vega itself, and this is about eighty times the distance between the Earth and the Sun, so that the quantity of material responsible was far from negligible; it was estimated to be as massive as all the planets in our Solar System put together.

61

It was not easy to make any estimate of the sizes of the particles moving round Vega, but Aumann and Gillett reasoned that very small dust-grains would long since have been drawn back into the star, leaving intermediate and large-scale debris in orbit. The possibility of a true planetary system began to emerge. The temperature of the material fitted in with this idea; it was given as about −184°C, which is about the same as that of the icy inner rings of Saturn.

There seems little doubt that the Sun's planets, including the Earth, grew up by accretion from a cloud of material associated with the youthful Sun itself. What can happen here can presumably happen elsewhere, and Vega, like the Sun, is a very commonplace sort of star. Gillett commented that the material could be in the form of 'a number of rings, in the stages of coalescing and possibly forming planets', but he was quick to add that this was really nothing more than controlled speculation. The next step was to see whether any other bright stars showed similar infra-red excess.

Some of them did − notably Fomalhaut (Fig. 24) in the constellation of Piscis Australis, the Southern Fish, which is only 22 light-years away, less than half the distance of Vega. Fomalhaut is well south of the celestial equator, and from England it is always very low down, while from Scotland it is difficult to see at all, but it is really quite a bright star; its magnitude is 1.2, about the same as that of Deneb. Like Vega, it has an A-type spectrum, and it is thirteen times as luminous as the Sun, with an estimated diameter of 1.7 million miles (about 2.75 million km).

Fomalhaut showed an infra-red excess comparable with Vega's. Other stars, such as Altair, did not. Further IRAS results indicated that there were some forty stars which could be associated with cool surrounding material, and the most significant case of all − though it did not appear so at the time − was that of a completely undistinguished star, Beta Pictoris in the far south (Fig. 25).

Pictor, the Painter, is not one of the original constellations, and was first drawn on the maps by the French astronomer Lacaille in 1752. It lies close to Canopus (Fig. 26), and can never be seen from England or the northern United States, because it does not rise above the horizon. It contains no bright stars, and to be honest there seems little reason for its existence as a separate constellation. Its only prior claim to fame was that in 1925 a bright nova flared up there, and reached the first magnitude before fading away.

Beta Pictoris, the second star of the constellation, is of magnitude 3.9. It has no individual name − it has never been regarded as worthy of one − and it is white, with a spectrum of type A5. The distance in the *Sky Catalogue* is given as 78 light-years, in which case Beta Pictoris is approximately sixty times as luminous as the Sun,

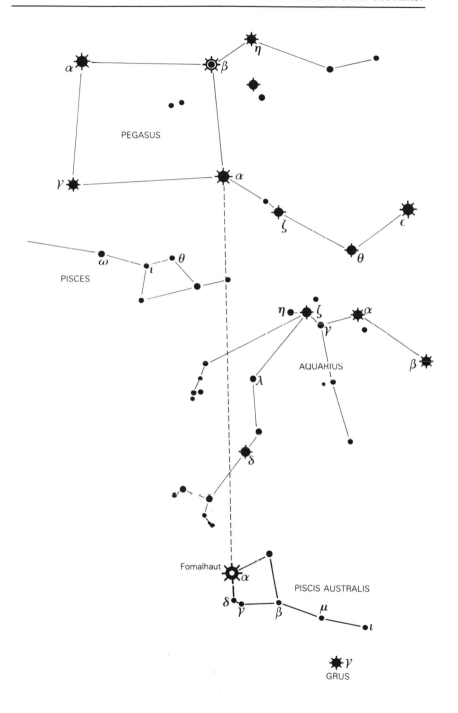

Fig. 24 Position of Fomalhaut

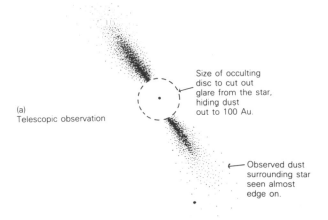

(a)
Telescopic observation

Size of occulting disc to cut out glare from the star, hiding dust out to 100 Au.

Observed dust surrounding star seen almost edge on.

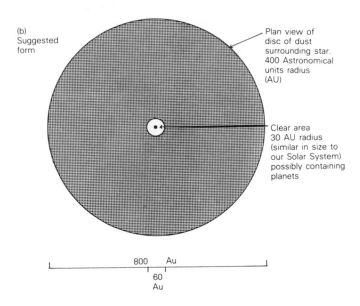

(b)
Suggested form

Plan view of disc of dust surrounding star. 400 Astronomical units radius (AU)

Clear area 30 AU radius (similar in size to our Solar System) possibly containing planets

800 Au

60 Au

Fig. 25 Material associated with Beta Pictoris

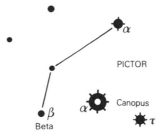

Fig. 26 Position of Beta Pictoris

making it slightly more powerful than Vega.

With the IRAS telescope, Beta Pictoris was found to have a quite exceptional infra-red excess. In fact, it was so marked that two American astronomers, Dr Bradford Smith and Dr Richard Terrile, set out to record the cool material visually. They used the 100-inch reflector at the Las Campanas Observatory, in Chile, which is of the same size as the 100-inch telescope at Mount Wilson, now sadly put out of commission because of the light pollution from nearby Los Angeles.

There are no large cities near Las Campanas, and conditions there are extremely good. Moreover, the telescope is used with electronic equipment. We are now in the midst of what I have called the 'electronic revolution', and the photographic plate is being super-seded, just as the camera replaced the human eye a century ago. Of special importance is the CCD, or charge-coupled device. It is far more sensitive than any plate, and when used together with a giant telescope it is immensely powerful.

With a CCD and the Las Campanas telescope, Smith and Terrile were successful almost at once, and recorded a disc of material which extended for nearly 50,000 million miles from Beta Pictoris in opposite directions from the star. The disc is seen nearly edge-on, and may be no more than a few hundred million years old. We can even make a shrewd estimate of the composition of the disc. Ices, silicates and carbon-rich substances are strong candidates, and these are the very materials from which the planets in our Solar System were formed. Analysis of the density of the Beta Pictoris material carried out by Smith and Terrile indicates that planets may have been produced there too, and that the innermost particles of the disc have already been swept away, perhaps by orbiting planets.

The real importance of this is that for the first time we have actually observed what may be another solar system. Up to now we have had to rely either upon infra-red results, or upon the admittedly very uncertain measurements of nearby stars which 'wobble' slightly as they are perturbed by companion bodies which are not massive enough to be stars. There is no doubt at all that the Beta Pictoris shell exists, and presumably the same is true of Vega, Fomalhaut and the rest.

Not everybody agrees on details. Re-analysis by David Diner and John Appleby, of the Jet Propulsion Laboratory, suggests that the disc of dust goes right up to Beta Pictoris and does not thin out near the star as Smith and Terrile claimed. This would make the chance of finding planets rather less likely. But we cannot be sure either way, and at least the possibilities are there.

This is not to suggest that there may be life on planets in these other systems – even if the systems themselves exist. We are not even sure of the origin of life on Earth. The general consensus of opinion is that it began in the seas at an early stage of the Earth's story, and certainly it was slow to evolve. Stars of type A, such as Vega, Fomalhaut and Beta Pictoris, are more luminous and more energetic than the Sun, so that they are running through their careers at a much faster rate. Any planets associated with them are likely to be much younger than the Earth, so that in all probability advanced life-forms would not have had time to develop.

It is all very tantalizing, but at least we have made a start. Vega, the first star to be found to show a strong infra-red excess, has already taught us a great deal – and certainly it gave astronomers a major shock when the telescope carried on IRAS found that it, like the Sun, may just possibly be the centre of a family of worlds.

VII ALGOL: THE DEMON STAR

There are a few stars which have acquired evil reputations. Of these, much the most notorious is Algol or Beta Persei, known as the Demon (Fig. 27). In the sky it marks the head of the Gorgon, Medusa, a hideous monster with a woman's face and body, but with snakes instead of hair, and a glance which could turn any living creature to stone. The Perseus legend is very much a part of Algol's story; many people will know it, but it may be worth re-telling here, albeit in very truncated form.

The innocent cause of the trouble was Andromeda, daughter of King Cepheus and Queen Cassiopeia, who were fairly conventional mythological rulers. Cassiopeia boasted that Andromeda was more beautiful than the sea-nymphs. This could well have been true; but as these particular nymphs were the children of the peevish sea-god, Neptune, the claim was a tactless one, and Neptune was irritated enough to send a sea monster to ravage the kingdom. Cepheus and Cassiopeia were in despair. They consulted the Oracle, and were told that the only way to appease the sea-god was to chain Andromeda to a rock by the water's edge, and leave her there to be gobbled up by the monster.

Luckily the gallant hero Perseus arrived on the scene in the nick of time. He had been on an expedition to kill the Gorgon, and was returning home, using winged sandals which had been kindly loaned to him by Mercury. He carried a shield, in which he had been able to see the reflection of Medusa's face without looking at it directly, and he now carried the severed head, carefully stowed away in a bag. On catching sight of the hapless princess tethered to the rock, he swooped down, opened the bag, turned the approaching sea-monster to stone and then, in the best traditions, married Andromeda. This is one of the few legends to have a happy ending, and all the main characters are on view during night-time: Cepheus

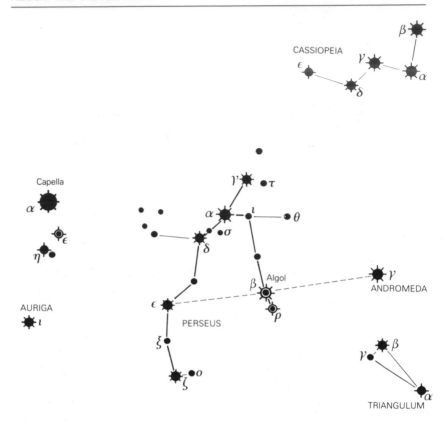

Fig. 27 Position of Algol

and Cassiopeia in the far north, Andromeda leading away from the Square of Pegasus, the monster represented by Cetus. Perseus is very prominent, and marking the Gorgon's head is Algol.

At first glance Algol looks like a perfectly normal star of the second magnitude, very slightly fainter than the Pole Star and considerably fainter than Mirphak, the leading star of Perseus; according to careful measurements Mirphak is of magnitude 1.8, Polaris 2.0 and Algol 2.3. The Arabs called it Ras al Ghul, the Demon's Head; later writers called it Caput Larvæ, the Spectre's Head; to the Chinese it was Tseih She, or the Piled-up Corpses. To astrologers it was the most unfortunate and dangerous star in the entire heavens.

In 1669 an Italian astronomer, Geminiano Montanari, noticed that there is something odd about it. It is not always at its brightest; sometimes it dips down by more than a magnitude. Subsequently, this periodical fading was confirmed by another Italian, Maraldi, and by a Saxon amateur named Palitzsch, who is best remembered as being the first man to sight Halley's Comet at its return of 1758 (the first time that the return of a comet had been predicted – but that is another story). However, the main investigations about Algol were carried out in 1782 by John Goodricke, one of the most unusual astronomers in history.

Goodricke was born in 1765, at Gröningen in Holland, though his parents were English and they soon came back to live in York. John was deaf and dumb. He never learned how to speak, but there was nothing the matter with his brain or eyesight, and he became fascinated by astronomy at an early age. He was an excellent observer, and he realized that the behaviour of Algol is predictable. Normally the star remains at maximum; but every two and a half days it begins to fade (Fig. 28), dropping to below the third magnitude in just over four hours. It remains at minimum for a mere twenty minutes, after which it slowly brightens up again, not to change noticeably for another two and a half days. The precise period between successive minima is 2 days 20 hours 48 minutes 56 seconds, and the value which Goodricke gave was remarkably correct. He also put forward the correct explanation. Algol, he said, is not truly variable at all. It is made up of two stars, one rather brighter than the other, and their revolution period is two and a half days. When the fainter star passes in front of the brighter, and hides or eclipses it, Algol gives its long, gradual 'wink'. It is better termed an eclipsing binary than a genuine variable star.

This was a flash of genius on Goodricke's part, and it was not the only one. In 1784 he discovered that two more naked-eye stars, Delta Cephei and Beta Lyræ, are variable. Delta Cephei is an intrinsic variable, while Beta Lyræ is an eclipsing binary, of a type rather unlike Algol.

How much Goodricke would have achieved will never be known, because he was not given the opportunity. In 1786 he died at his home in York. He was only twenty-one years old, and provides one of the most notable cases of a brilliant man who overcame what most people would have regarded as a hopeless handicap – only to have his career cut tragically short. He is not forgotten, particularly in York, where there is a Goodricke Society devoted to helping and caring for children who are born deaf and dumb.

Algol, the Demon Star, lies in the Gorgon's head. Surely, then, the ancients must have known about its 'winking'? This seems logical enough but, curiously, there is no evidence. One of the best of the Arab star catalogues was drawn up by Al-Sufi in the tenth century, but Al-Sufi does not mention Algol's fluctuations, and the general opinion among modern scholars is that the discovery by Montanari in 1669 really was the first intimation that Algol was not constant in light.

69

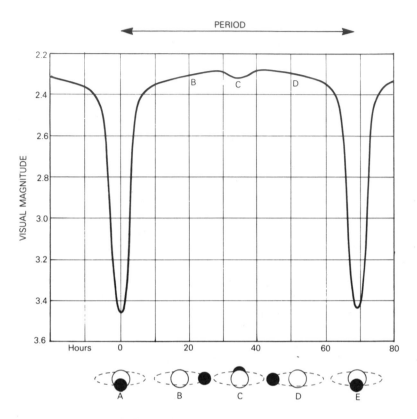

Fig. 28 Light-curve of Algol

Algol is the closest and therefore the brightest of all eclipsing binaries; it is 94 light-years away. Other systems of the same type, visible with the naked eye, are Lambda Tauri in the Bull, and Delta Libræ in the dim Zodiacal constellation of the Balance. Beta Lyræ is also an eclipsing binary, but its behaviour is different from Algol's. In the far south, never visible from Europe, we have another Algol star, Zeta Phœnicis, with a range of from 3.6 to 4.1, and a period of 1.7 days.

Careful measurements of the period of Algol were made in 1855 by a famous German astronomer, Friedrich Wilhelm Argelander, who is best remembered for his great star catalogue – the *Bonner Dürchmusterung* – which includes no less than 324,198 stars. Argelander found that Algol's period had shortened by 6 seconds since the time when Goodricke had made his observations. Subsequently it was established that the period changes regularly in a cycle of about 680 days, and Argelander realized that this must be due to the variation in distance which the light from the bright star takes to reach us – because it and its eclipsing companion are themselves in motion round a common centre of gravity with a third star, now called Algol C. Much more recently Algol C has been tracked down spectroscopically, and there is also some evidence of a fourth component, Algol D.

In fact, there was no absolute proof of the truth of Goodricke's theory until 1889, when some pioneer work was undertaken by the director of the Potsdam Observatory, Hermann Carl Vogel. Vogel carried out a survey of the spectra of over four thousand stars, and he was particularly interested in Algol, because the lines in the spectrum oscillated to and fro around a mean position. He concluded, rightly, that this was due to the Doppler effect. Algol really did consist of two components, but the spectrum of the fainter member (Algol B) was too faint for him to detect, and all he could see was the shift in position of the lines due to the brighter star as it moved round in orbit. It was not until 1978, almost two centuries after Goodricke's time, that the elusive spectrum of Algol B was tracked down. The light-curve (Fig. 28) shows that there is a very slight minimum when the brighter star passes in front of the fainter; the drop is only about a tenth of a magnitude, so that it is virtually impossible to detect with the naked eye.

All these various investigations lead us on to a picture of the Algol system which may not be complete, but which seems to be at least fairly reliable – and this is of great importance, because a knowledge of the movements and sizes of the stars in a binary make it possible to determine their masses.

The main component (Algol A) is of spectral type B, and is a white star, around a hundred times as luminous as the Sun. Its diameter is roughly 2.6 million miles (about 4 million km). Algol B, the eclipsing companion, is of type G; it is not genuinely dark, and appears to be three times as luminous as the Sun. It has a diameter of about 3.5 million miles (about 5.5 million km), so that it is larger than the bright star and qualifies as a sub-giant, but its mass is less than that of A, which seems at first sight rather curious. According to the spectral types, it should be the more massive of the two.

The explanation is bound up with the way in which a star is born, and how it evolves. As we have seen, the gas-and-dust clouds known as nebulæ are the stellar nurseries, and we may be fairly sure that the two components of a binary such as Algol were formed out of the same nebula at the same time. The more massive star evolved the more quickly of the two, and left the Main Sequence earlier, moving into the giant branch of the HR diagram. Yet with Algol, it is the less evolved star (A), still on the Main Sequence, which has the greater mass.

It was Sir Fred Hoyle who first gave a reason for this. What seems to happen is that the G-type component was originally more massive than its partner, so that it left the Main Sequence at an earlier stage and began to swell out. As it increased in size, its gravitational grip on its outer layers weakened, so that material from these layers was 'captured' by the companion. Eventually the situation was reversed; the star which was originally the less massive became the senior partner. The process is still going on, as we can tell from long-wavelength observations; the Algol system is a radio source, indicating that a stream of material is making its way from B to A. This is what is called mass-transfer, and is of tremendous importance in all studies of binary systems.

With Algol, the eclipses are not total, so that even at minimum light there is still a little of the bright component left showing. If we were observing the system from a different direction, the eclipses might be total – but from other vantage points there would be no eclipses at all, and Algol would appear constant. All eclipsing stars are close binaries, but from our position in the Galaxy by no means all close binaries are eclipsing pairs. For example, Mizar A is not.

Because Algol is so bright, its fluctuations are easy to follow with the naked eye. The procedure is to compare it with nearby stars which do not change, and which can be used as comparisons (Fig. 29); the magnitudes of some comparison stars are 1.8 for Mirphak or Alpha Persei, 2.1 for Gamma Andromedæ, 2.8 for Zeta Persei, 2.9 for Epsilon Persei and 3.8 for Nu Persei.* Times of minima are given in various journals and almanacs, and because they are so frequent it is easy to pinpoint them. If you look up at Perseus and find that the 'Demon' is not so prominent as usual, you may be sure that an eclipse is in progress.

The other eclipsing binary observed by John Goodricke is of a different type, and in its way is even more remarkable. It lies close to Vega, and it has a name of its own – Sheliak – but most people prefer to call it by its official title of Beta Lyræ.

Lyra is a small but interesting constellation (Fig. 30). Vega, of course, dominates it, but there are several other features of special note. There is, for example, the system of Epsilon Lyræ. Keen-sighted people can see that Epsilon is double; a telescope shows that each component is itself made up of two, so that we have a double-double, or quadruple, system. Also in the constellation is a red variable, R Lyræ, and a much fainter variable which is called, rather confusingly, RR Lyræ; I will have more to say about it later. There is, too, the most famous of all planetary nebulæ, Messier 57 (the

---

*Avoid using the red star Rho Persei, which is itself variable between magnitudes 3 and 4, but with no constant period. It is not an eclipsing binary, but is of the same basic type at Betelgeux, though it is not nearly so powerful or remote.

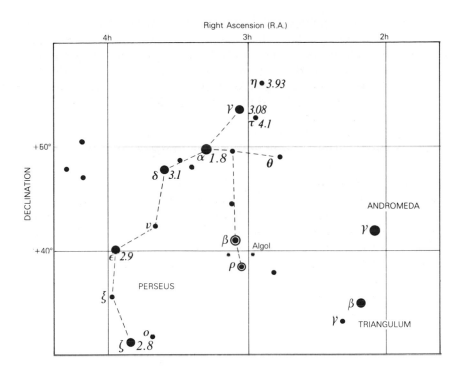

Fig. 29 Chart of comparison stars for Algol

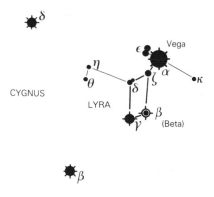

Fig. 30 Position of Beta Lyræ

57th object in a catalogue of clusters and nebulæ drawn up by the French astronomer, Charles Messier, in 1781). M57 is much too faint to be seen with the naked eye, and I have not been able to glimpse it with binoculars, though apparently some people can do so. Telescopically it is revealed as a dim, luminous ring, in the centre of which is a faint star. Our Sun will certainly become a planetary nebula after its red giant stage.

Two stars flank M57 to either side. One is Gamma Lyræ, of magnitude 3.2. The other is our eclipsing binary, Beta.

This time the period is almost 13 days, much longer than Algol's, and there are alternate deep and shallow minima; the variations are always going on (Fig. 31). The maximum magnitude is just below 3.4, so that the star is slightly fainter than its neighbour Gamma. Beta then fades down to magnitude 3.8, recovers, and then goes through its primary minimum, which takes it down to below 4, after which it returns to maximum and the entire cycle starts once more.

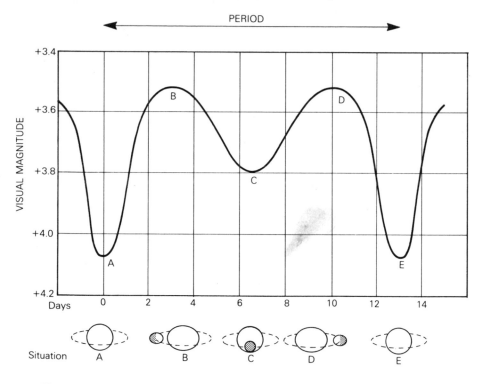

Fig. 31 Light-curve of Beta Lyræ

Gamma is the best comparison star; you can also use the nearby Delta and Kappa (each 4.3).

We cannot see the components of Beta Lyræ separately, because it seems that they are almost touching one another. They are pulling upon each other, and are drawn out into egg-like shapes; it has even been said that they behave rather in the manner of two boiled eggs rolling around end-to-end. When they are broadside-on to us, they appear brighter than when they are end-on, which makes the light-curve very complicated. In addition there are reflection effects, more marked than with Algol, because each star shines upon the other and increases the luminosity of the nearest part of its globe.

Beta Lyræ is much more remote than Algol. One estimate gives its distance as 300 light-years, but this may be too low; in any case, both components are much more powerful than the Sun. The 'atmospheres' of the two must touch each other, and there is evidence that the stars are connected by huge streamers of gas which move at about 200 miles per second (320 km per sec.). The primary minimum as seen from Earth is caused by a total eclipse, the secondary minimum by a partial eclipse.

One surprising fact is that the fainter component is the more massive of the two, so that it is far less luminous than it ought to be. One possible solution to this puzzle is that much of the star's mass is contained in a cloud or disc of material which surrounds the main body, and it may even be that this cloud envelops both components, in which case the spectacle from close range would be incredible. Meanwhile, the fainter, more massive star is continuing to draw matter away from its companion, but this will not continue indefinitely. When the secondary swells out, the process will be reversed, and matter will be re-captured by the primary until it in turn becomes dominant. It is rather a case of stellar 'tit for tat'.

Plenty of eclipsing binaries are now known, some similar to Algol, others of the Beta Lyræ type, and also dwarf pairs known as W Ursæ Majoris stars, where the components are feeble dwarfs and are very close together. But it was Algol which gave the first key to what is going on, and it is always worth looking at. To the ancients it may have been the fearsome Demon, but to modern astronomers it is a source of never-ending interest.

VIII EPSILON AURIGÆ: THE MYSTERIOUS 'KID'

Look close to Capella, the brilliant yellow star which is almost overhead during winter evenings in England, and you will see a triangle made up of three much fainter stars, which are lettered Epsilon, Zeta and Eta Aurigæ (Fig. 32). Eta is a completely normal B-type star, 200 light-years away and with a luminosity 400 times that of the Sun. The other two are extraordinary by any standards. Together the three have been nicknamed the 'Hædi', or Kids, because Capella was once known as 'the Little She-goat'.

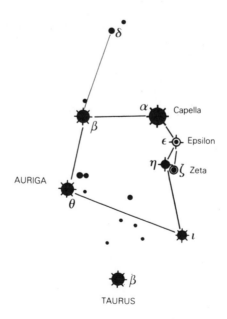

Fig. 32 Positions of Epsilon and Zeta Aurigæ

The apex or northern star of the triangle, Epsilon Aurigæ, has never been dignified with an accepted proper name. There are hints that the Arabs called it Almaaz or Alanz, but the name is never used today – which is rather a pity, because the star is much too important to be relegated to comparative anonymity. In some respects it is the most puzzling of all the naked-eye stars, and even today we cannot claim that we have a clear picture of what it is like.

Its usual magnitude is 3.0, so that it is slightly brighter than Eta, but as long ago as 1821 a German astronomer named Fritsch announced that it is variable. He seems to have made no further comments about it, but in 1847–8 the variability was fully confirmed by three more German observers, Argelander, Heis and Schmidt. Julius Schmidt (best remembered for his excellent map of the Moon) made further studies of the star in 1874–5, and was in no doubt about its fluctuations. In 1890 G. F. Chambers, a London barrister who was an ardent amateur astronomer, gave a catalogue of variable stars in which Epsilon Aurigæ was listed as irregular, with a range of from magnitude 3 to 4.5. He added that according to some observers, the variations were imperceptible over long periods, and that there was no regularity about them.

There matters rested until 1912, when H. Ludendorff, one of the leading experts on variable stars, decided to make an analysis of all the past observations of Epsilon Aurigæ. What he found was extraordinarily interesting, and quite unexpected. There were three periods when the star had been recorded as fainter than usual – in 1821, 1847–8, and 1874–5. Ludendorff came to the conclusion that the star was not genuinely variable at all. It was an eclipsing binary.

This was something new. Of course many Algol and Beta Lyræ stars were known by then, but the periods were always short, amounting to a few days or even a week or two. With Epsilon Aurigæ the period was 27 years, and the eclipse lasted for many months instead of only twenty minutes for Algol. Also, where was the eclipsing component – the secondary? There was no trace of it, either visually or spectroscopically.

The changes in brightness were not very marked; from the usual maximum of 3.0, Epsilon Aurigæ fell to only about the fourth magnitude at minimum, which meant either that the eclipses were not total or that the secondary was quite bright on its own account. The next step was to try to find out the distance of the system. This was no easy matter, because Epsilon Aurigæ is much too remote to show any reliable parallax, and the only method was to study the spectrum of the primary component – which turned out to be of type F0, indicating that the star was a very luminous supergiant. The best modern estimate of the distance is rather over 4000 light-years. If this is correct, the primary must be equal to about 200,000

Suns, far outshining even Deneb or Rigel. Even if the value of 4000 light-years is too high, Epsilon Aurigæ is still extremely powerful. If it were as close to us as (say) Sirius, it would rival the Moon.

The mystery of the secondary component remained. But for the fact that it eclipsed the primary every 27 years (to be more precise, every 9883 days) it would be absolutely unknown, and there would be no reason to regard Epsilon Aurigæ as anything more than a remote supergiant; it would not even hold the 'luminosity record', since we know of at least one star, Eta Carinæ, which can match several million Suns.

Something could be learned about the masses of the two components. From the way in which the primary moves in relation to the invisible secondary – and here, of course, we come back to our reliable ally, the Doppler Effect – it seemed that the main star is as massive as thirty-five Suns, while the secondary must have about twenty solar masses. But in a way, this increases the problem. A star with as high a mass as the secondary ought to be detectable, either in infra-red (if very cool) or at least by its spectrum, but nothing at all could be found. The secondary component was as elusive as any ghost. Theorists cast around for an answer, and produced several, none of which seemed to be entirely satisfactory.

The first was proposed in 1937 by several astronomers in America (Gerard Kuiper, Bengt Strømgren and Otto Struve). They assumed that the secondary was a very young star, which was still condensing out of interstellar material and had not yet become hot enough to shine. It would, in fact, be a 'star in the making'.

At first sight this seemed reasonable enough. We are virtually certain that all stars do begin their lives in this way, and that they cannot emit visible light until their temperature has become sufficiently high. Unfortunately there were some disquieting facts. For example, infra-red radiation from a star in the process of formation should make its presence known well before visible light, and there was absolutely no sign of infra-red from the Epsilon Aurigæ secondary. Moreover, even at mid-eclipse more than half the normal light of the system still reaches us, and it is not easy to believe in a large, invisible star which is also transparent. Neither does the primary show any colour change when eclipsed, whereas material moving between the star and ourselves would presumably cause reddening.

Calculations were made to see how big the secondary would have to be. From the behaviour of the primary, and the duration of the eclipse – more than two years from start to finish – it was clear that the diameter of the secondary would have to be more than 2000 million miles (3200 million km), large enough to swallow up the orbits of all the planets in our Solar System out to beyond Saturn

(Fig. 33). There were serious doubts as to whether an embryo star of this size could be stable, because it would be comparatively close to its powerful, massive companion, and would be subject to enormous tidal forces. It would, incidentally, be almost unbelievably rarefied, with an average density of no more than a thousand-millionth that of the Sun. All things considered, the theory began to look decidedly far-fetched.

The next idea was different. It was suggested that the secondary was not an ordinary star at all, but a flattened, disc-shaped mass which periodically moved edge-on to the primary, as seen from Earth, so that the primary would be bisected – as shown in Fig. 34. Parts of the primary would remain unobscured, which would at least explain why the light we receive drops by no more than a magnitude at any time. Yet once again we come face to face with problems of instability: could such a disc survive so close to a star with thirty-five times the Sun's mass? And why is there no detectable increase in temperature toward the centre of the agglomeration, as we would expect to be able to find spectro-

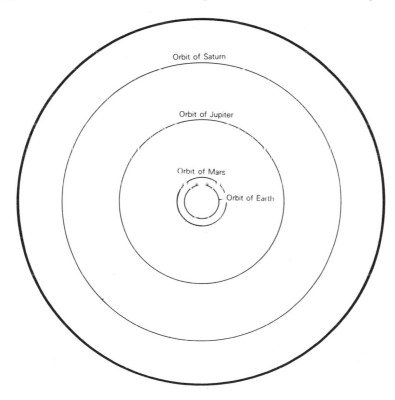

Orbit of Saturn

Orbit of Jupiter

Orbit of Mars

Orbit of Earth

Fig. 33 Size of Epsilon Aurigæ according to an early theory

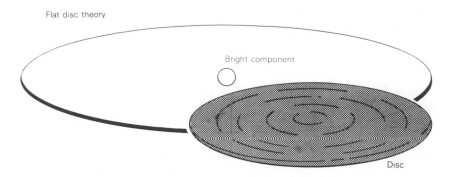

Fig. 34 Flattened disc theory of Epsilon Aurigæ

scopically? Rather reluctantly the whole idea was cast aside, and theorists looked around for something better. They found it in the concept of a black hole.

To give some idea of black holes (assuming that they exist at all, which is not absolutely certain), we must go back to what I have already said about the life-stories of the stars. The Sun will turn into a white dwarf, as the companion of Sirius has already done. A more massive star will suffer a spectacular fate; it will explode as a supernova, and then be transformed into a very small, incredibly dense object together with a cloud of expanding gas. Logically, I suppose, it would be right to talk about supernovæ before black holes, but I propose to defer them until Chapter 11, because they certainly do not come into any discussion of Epsilon Aurigæ.

Consider, then, a very massive star, much 'heavier' than the Sun. It will run through its evolution quickly by cosmic standards, and will not shine steadily for anything like the Sun's placid period of around 10,000 million years on the Main Sequence. It will be spendthrift of its nuclear fuel, and quite suddenly the crisis will come: when the production of energy stops, the star begins to collapse, and gravity takes over.

Gravity, remember, is the most important force in the entire universe, and with our very massive, exhausted star it becomes dominant. The star goes on collapsing and collapsing, becoming smaller and smaller as well as denser and denser, and as this happens the escape velocity increases. Most people today are familiar with the term escape velocity; it is the speed needed to break free from a body (star, planet or anything else) without any extra impetus. With the Earth, it amounts to 7 miles per second (about 11 km per sec.). Launch a missile at this speed, and it will never come back, but will simply move away into outer space, whereas if it departs at a lesser velocity the Earth's gravity will be able to hold it. (I admit that this is very much of an over-simplification, but it will do for the moment.) If the Earth were more massive, the escape velocity would be greater; it is 37 miles per second (59.2 km per sec.) for Jupiter, and as much as 400 miles per second (640 km per sec.) for the Sun.

But escape velocity also depends upon the size of the body: the smaller the size, the higher the critical speed. So with a collapsing star, the escape velocity grows and grows as the star becomes smaller. Finally we reach a stage where the escape velocity has soared to 186,000 miles per second (300,000 km per sec.) – and this, of course, is the velocity of light. Not even light can now break free; and if light cannot do so, then certainly nothing else can, because light is the fastest thing in the universe.

We have a strange situation. The old star, sometimes termed a 'collapsar', has surrounded itself with a region where the gravitational tug is so strong that nothing whatsoever can escape. This is what is called a black hole. You can enter a black hole, but you can never leave it; the region inside the boundary – the so-called event horizon – is to all intents and purposes cut off. What happens there is something which we cannot visualize, because all the ordinary laws of science break down. It has even been suggested that the old star may crush itself out of existence altogether. There are also some frankly peculiar theories about using black holes as means of transport to another part of the universe, or into a different universe altogether. To be candid I am highly sceptical about this sort of idea; but when asked to say what the conditions inside a black hole are, I am completely at a loss.

There are some eminent astronomers who question whether black holes exist at all, but there is fairly strong evidence that they do. Probably the best candidate is Cygnus X1, in the Swan, which consists of a supergiant star (HDE 226868) attended by an invisible companion which may well be a black hole. Obviously we cannot see it, but the system is a powerful source of X-rays, which can be produced only by intensely heated material. The scheme is that the black hole is pulling material away from the supergiant, and that before the captured material is drawn over the event horizon, to vanish forever from our ken, it is so strongly heated that it sends out the X-rays which we receive. The material has formed what we call an accretion disc round the black hole.

There are plenty of X-ray sources in the sky, but until the Space Age we knew very little about them, because we cannot study them from the Earth's surface; the rays are absorbed by the atmosphere. It was only in 1962 that the first rocket carrying X-ray detectors flew above the blanket of air and sent back reliable information. Today we obtain most of our X-ray data from artificial satellites, which

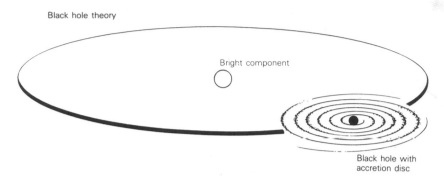

Fig. 35 Black hole theory of Epsilon Aurigæ

have been very successful. Without space-research methods, we would still be blissfully unaware of the short-wave radiation from Cygnus X1.

Very well, then – suppose that the invisible secondary of Epsilon Aurigæ is a black hole? This would explain why it could not be detected visually or spectroscopically, and if surrounded by an accretion disc of sufficient size it could well cause regular eclipses of the primary (Fig. 35). For some years the black hole picture was popular – and it must be added that during that period, which lasted through most of the 1970s, there were some theorists who tended to regard black holes as convenient explanations for any phenomena which could not be understood otherwise. Alas, there are serious drawbacks. For one thing, we have seen that Cygnus X1, the most plausible black hole candidate, is an exceptionally strong X-ray source; Epsilon Aurigæ is not, and in fact there is no trace of any X-ray emission at all. Moreover, it is hard to see how a vast accretion disc could remain uniformly cool and dark so close to a powerful supergiant which has a tremendous pull of gravity.

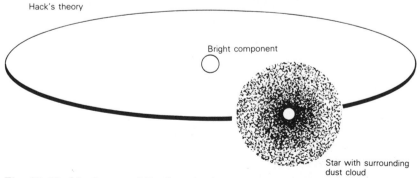

Fig. 36 Hack's theory of Epsilon Aurigæ

Gradually the black hole theory for Epsilon Aurigæ fell from favour, and opinion started to swing back to an idea proposed as long ago as 1961 by the Italian astronomer Margarita Hack. This time we have no infant star, no unlikely-sounding accretion disc, and no black hole. Instead, we have a fairly normal-sized, very hot blue star which is contained inside a huge shell of gas, no doubt with dust as well (Fig. 36). It is this shell, so strongly heated by the central star that it has become opaque, which causes the eclipses. We have avoided the difficulty of assuming that there must be a star big enough to engulf much of the Solar System, and we can also explain why the secondary component cannot be observed: its light is completely blocked out.

If the shell of material round the secondary is being strongly heated, it would be expected to emit short-wave radiation – not in the X-ray region, but in the rather less extreme ultra-violet. Recent results from the successful satellite IUE (International Ultra-violet Explorer) seem to indicate that there is some radiation of this kind from Epsilon Aurigæ. The evidence is not conclusive, but it is a pointer, and by now most people are fairly happy about accepting Margarita Hack's explanation.

Of course, it too has its drawbacks. Some hot blue stars are known to be associated with shells which they have ejected, but all of these stars seem to be less massive than Epsilon Aurigæ by a factor of about a thousand, and it is not easy to see just how so tremendous a shell could be formed. All we can really say is that the theory is less unlikely than the others, but we certainly cannot yet claim that we know what Epsilon Aurigæ is really like.

It would be a great help if we could detect other systems of the same kind. The closest parallel is probably a faint eclipsing binary in the far north of the sky, VV Cephei, which is never bright enough to be seen with the naked eye, not because it is intrinsically feeble but because it is so remote. It, too, has a long period, 20.3 years (7430 days), and it, too, has a supergiant primary, but the spectral type is M – the same as that of Betelgeux – and the spectrum of the secondary is visible; it is of type B9, so that the star is hot and bluish. Eclipses last for fifteen months, but at least we know the nature of the secondary, and there is no guarantee that it is really similar to the invisible component of Epsilon Aurigæ.

By sheer coincidence, the other important eclipsing binary of long period is another of the Kids: Zeta Aurigæ, the dimmest member of the triangle, sometimes still known by its old proper name of Sadatoni. Here the period is 2.66 years (972 days), and again we have a supergiant primary with a smaller, hotter companion. It is when the hot star is hidden by the supergiant that we see a drop in light, but the visual variations are by no means obvious, because the range is small. At its brightest the magnitude is 3.7, and it never drops below 4.2.

There is only a rough analogy with the brighter Kid, because with Zeta Aurigæ we can not only see the spectra of both components, but can even watch what happens when the eclipse begins (Fig. 37). As the supergiant starts to pass in front of the hot star, the light of the secondary comes to us through the giant's outer layers, and there are complicated changes which are highly informative. Gradually the eclipse becomes more and more marked, and finally becomes total, remaining so for 38 days before the observed spectral changes take place in the reverse order.

We have learned a great deal about Zeta Aurigæ, but we are not sure of its distance, and estimates range from 500 light-years out to as much as 1200 light-years. However, we can be sure that the system is much closer and much less luminous than Epsilon. At most the primary can hardly be more than 2500 times as powerful as the Sun, and is probably less, while the hot secondary could match no more than 400 Suns even if the system is as far away as some of the measurements indicate. It is pure chance that Epsilon and Zeta lie side by side in the sky; there is absolutely no connection between them.

I suppose that the main lesson to be learned from Epsilon Aurigæ is that there are many objects in the Galaxy that we cannot see. If we were observing from a different direction, there would be no eclipses of the primary of the system, and we would have no reason

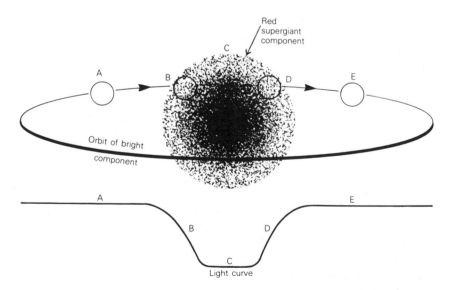

Fig. 37 Theory of Zeta Aurigæ

to suspect the existence of any companion – just as we would not have found the secondary in Algol without the fluctuations in light (or, at least, we would not have found it so soon). And if Epsilon Aurigæ is attended by a massive invisible body, why should not other normal-looking stars be similar? There is no reason to suppose otherwise.

Youthful star, flattened disc, black hole, or hot star with a surrounding shell – it is not really fair to say 'Take your choice', because the last of these explanations has so much to offer, but Epsilon Aurigæ has already given us so many surprises that we must be prepared for more. We may obtain new information from the Space Telescope, named in honour of the great American astronomer, Edwin Hubble, which was due to be launched in 1986 but which was postponed because of the disaster with the *Challenger* shuttle. The telescope will have a 94-inch mirror, and since it will be operating from above the top of the atmosphere it will be immune to the irritating absorbing effects which so hamper us at ground level. No doubt it will be directed toward Epsilon Aurigæ, and we may then be able to say definitely whether there is a chance of detecting the secondary at any wavelength.

Meanwhile, I fear that visual observers will have a long wait before anything else happens in the Epsilon Aurigæ system. The last eclipse began on 22 July 1982; it became total on 11 January 1983; totality ended on 16 January 1984, and the partial phase came to an end on 25 June 1984, so that the next eclipse is not due until 2011. On the last occasion, I made regular estimates with the naked eye, comparing Epsilon with its neighbour, and I was easily able to see how the light faded from the usual magnitude of 3 down to minimum in the middle of 1983 (mid-eclipse actually occurred on 12 July). The light curve given here (Fig. 38) is drawn from these estimates, and does not pretend to be particularly accurate, but at least it shows the sequence of events. However, there are suggestions that the magnitude is not quite constant even at times of non-eclipse, and it is always worth keeping a watch. If you find that Epsilon Aurigæ is much brighter or much fainter than it ought to be, then there will be something unusual going on – but do not expect anything of the kind, or you may be in for a disappointment.

Epsilon Aurigæ has a faint visual companion of magnitude 14, discovered way back in 1891. It may be a member of the system, but it is at least half a light-year away from the main pair, and in my view, at least, the connection is decidedly dubious. It is over 28 seconds of arc away from the visible star, so that moderate-sized telescopes will show it.

It is fascinating to speculate as to what Epsilon Aurigæ would look like from close range. There just could be a disc of gas and dust, or a

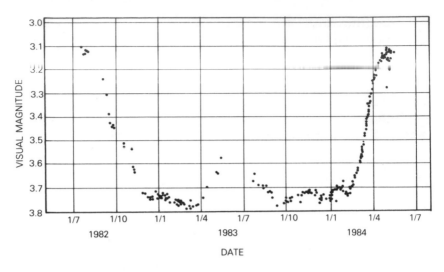

Fig. 38 Light-curve of Epsilon Aurigæ during the last eclipse

void indicating the presence of a black hole; much more probably there would be a tremendous, yellow sun accompanied by a cloud which concealed the hot secondary. The presence of any planets, inhabited or otherwise, is unlikely, but again we cannot be sure. With Epsilon Aurigæ, almost anything is possible. Next time the sky is clear, I suggest that you go outdoors, find Capella, locate the faint triangle nearby, and take a careful look at the remarkable 'Kid' which has set astronomers so many problems.

IX MIRA: THE 'WONDERFUL STAR'

On 13 August 1596, a Dutch amateur astronomer named David Fabricius was making some rather casual observations of the stars. He was deeply interested in the sky but he was by profession a minister of the Reform Church, and at various times held charges in Resterhave and Osteel. Among his friends (by correspondence at least) were the eccentric Dane, Tycho Brahe, and the German theorist and mystic, Johannes Kepler, who was the first to show that the planets move round the Sun in orbits which are elliptical rather than circular.

On that particular August night Fabricius was looking at the constellation of Cetus, the Whale, which is sometimes identified with the fearsome sea-monster of the Perseus legend (Fig. 39). Cetus is not a particularly striking group; its leading star, Diphda, is only of the second magnitude, and lies south of the Square of Pegasus (unwary observers have often confused it with Fomalhaut in the Southern Fish, though Diphda is a magnitude the fainter of the two and is rather higher up as seen from northern latitudes). The Whale's 'head' is quite easy to make out and contains one star, Menkar, which is obviously orange.

However, Fabricius' attention was concentrated upon a star of about the third magnitude, some distance from the Whale's head. It was not a star which he knew, though his knowledge of the sky was excellent. Outwardly there seemed nothing strange about it apart from its sudden arrival, but Fabricius noted that a few weeks later it was no longer to be seen. So far as we can tell, he thought no more about it, thereby missing the chance of making an important discovery.

Fabricius, was one of the early telescopic users, but his career was cut short in an unexpected manner. In May 1617 he preached a sermon in which he said he had lost a goose, stolen by one of

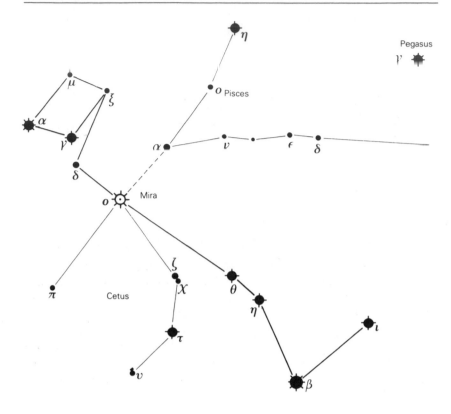

Fig. 39 Position of Mira

his parishioners, and he said that on the following Sunday he would reveal the name of the culprit. This was unwise, because before next Sunday came Fabricius had been murdered, no doubt by the goose thief. (His son, Johann, was also an astronomer and a promising telescopic observer, but had died a few weeks before at the early age of twenty-nine.)

The next record of the star in the Whale came in 1603, when Johann Bayer was compiling his famous star catalogue. You will remember that Bayer gave the stars their Greek letters, beginning with Alpha and working through to Omega. With Cetus he confused the order, as he so often did, so that the five brightest stars are, in order, Beta (Diphda), Alpha (Menkar), Eta, Gamma and Tau; but, more importantly, he re-observed Fabricius' star, and gave it the Greek letter Omicron. Once more it vanished after a few weeks, and again there was no serious attempt to follow it up; we are not sure whether Bayer knew of Fabricius' observation – very probably he did not, or his suspicions would have been aroused.

Next in the story of this unusual star came another Dutchman,

Johannes Phocylides Holwarda, who, unlike Fabricius, was a professional. He had been born in Friesland in 1618, and had become a professor at the University of Franeker. He wrote a book about astronomy, in which he made the suggestion that stars may have small individual or proper motions of their own, in which he was of course quite right, though it was to be another six decades before Edmond Halley proved it. Meanwhile, Phocylides – as he is always known – made many useful observations, and during a total eclipse of the Moon, in December 1638, he saw the puzzling star once more. Checking back in the records, he found that it had been previously seen not only by Fabricius and Bayer, but also by the German astronomer Wilhelm Schickard. To see three novæ in the same position would really be too much of a coincidence, and Phocylides realized that we must be dealing with a variable star which periodically brightened and faded. Then, in 1648, one of the leading observers of the time, Hevelius of Danzig, saw it again, and named it Mira, or 'the Wonderful'. It is still known as Mira today.

On average the period is 331 days, so that Mira comes to maximum about a month earlier every year. This means that there are several consecutive years in which maximum occurs when the star is above the horizon only during daylight, which is undoubtedly the main reason why nobody recognized it as a variable star until Phocylides did so. At its peak it can become very conspicuous. According to reports which are probably reliable, it rose to the first magnitude in 1779, and was nearly equal to Aldebaran. I have never seen it as brilliant as this, but in 1969 and again in 1987 it became practically as bright as Polaris; I made the magnitude 2.3. At other maxima it does not rise above the fourth magnitude, and occasionally there are maxima which barely reach naked-eye visibility. It is certainly too faint to be seen without optical aid for nine months out of the twelve, and usually for rather longer.

The magnitude at minimum was less easy to determine, at least by early telescopic observers. Halley stated that it was never 'totally extinguished, but may at all times be seen with a 6-foot tube' (that is to say, a telescope with a focal length of six feet). Not everyone agreed. In a well-known book about astronomy, published in 1865, Amédée Guillemin claimed that at minimum 'it becomes completely invisible, not to the naked eye only, but even in our telescopes . . . we are surprised that observations of this singular star have not been pursued with the most powerful telescopes during the period of invisibility. It is only known that it is then below the 11th magnitude.' In fact it never drops much below magnitude 10, and I can always find it easily with a modest 3-inch refractor, and it is not hard to find. It lies close to a pretty little double star, 66 Ceti, whose components are of magnitudes 5.7 and 7.7 and are separated by

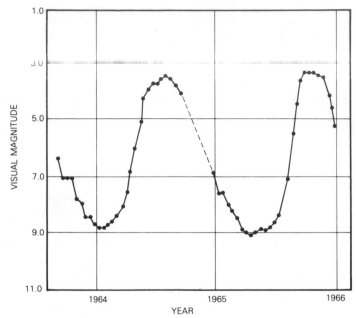

Fig. 40 Light-curve of Mira

over 16 seconds of arc. Note also that Mira lies in line with the three naked-eye stars Eta, Omicron and Alpha Piscium (see Fig. 39). If you use 7 × 50 binoculars, Mira is just in the same field with Delta Ceti, in the Whale's head. My advice is to find it when it is bright, so that you can see it easily with the naked eye, and then memorize the binocular and telescopic fields so that you can track it down again when it is faint.

Mira has a tenth-magnitude companion. The apparent separation is less than one second of arc (at present it is about 0".8) but the companion is not a difficult object when Mira itself is at minimum and the two members of the pair are equal in brightness. Their colours are not the same; Mira is orange-red, but the companion is hot and bluish, and appears to be what is termed a sub-dwarf – that is to say a star which is smaller but denser and more luminous than our Sun, intermediate in type between a true Main Sequence star and a white dwarf.

We can hardly doubt that Mira and its companion are associated, because they have the same motion through space and are moving away from us at the same rate (about 24 miles per second (38 km per sec.), as we can tell from the Doppler shifts in their spectra). They must be at least 5000 million miles apart (8000 million km), which is considerably more than the distance between the Sun and the outermost planets of the Solar System, Neptune and Pluto. There is

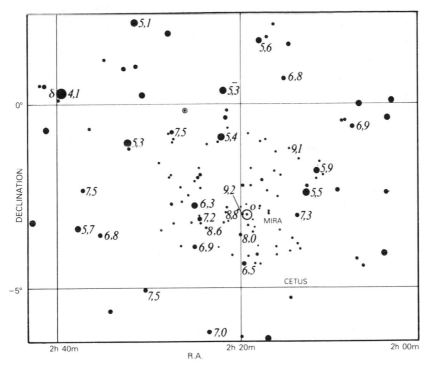

* Note: The numbers by the stars are their visual magnitudes

Fig. 41 Fields of Mira: comparison stars

some evidence that the companion is itself variable in light, but not to any great extent, and its fluctuations, even if real, are quite unlike those of Mira itself.

The distance of Mira is not known with precision. One recent estimate gives 220 light-years; the *Cambridge Sky Catalogue* gives 95 light-years. In any case, Mira is much closer than other stars of its kind, which is why it looks comparatively bright. It is far from unique. By now many thousands of similar stars are known, and we usually call them Mira variables, though unofficially they are known as long-period variables. Apart from Mira, the only examples which can attain more than the fringe of naked-eye visibility are R Carinæ in the southern constellation of the Keel (3.9), Chi Cygni in the Swan (3.3), R Horologii in the Clock (4.7) and R Hydræ in the Sea-Serpent (4.0). It is interesting to note that by the end of the seventeenth century only three stars were known to be variable – Mira, Algol and Chi Cygni – and only another nine were added before 1800.

Like all long-period variables, Mira is red, with an M-type

91

spectrum of the same basic type as that of Betelgeux. Neither its period nor its amplitude can be predicted with certainty. The period may be around a week longer or shorter than the mean of 331 days, and, as we have seen, the maximum magnitude has a considerable range over different cycles, though the minimum magnitude always appears to be about the same. We are certainly not dealing with an eclipsing binary, such as Algol or Epsilon Aurigæ. Mira is intrinsically variable; like Betelgeux it swells and shrinks, changing its output of radiation as it does so, though, unlike Betelgeux, the changes are very marked, and the period is at least roughly regular.

Mira is a huge star, with a diameter which may be as much as 500 times that of the Sun when it is at its largest. It is also extremely cool by stellar standards. The surface temperature may drop to only about 2000°C, and never rises as high as 3000°C, which explains why Mira is red. Oddly enough, the highest surface temperature is reached not when the star is at its brightest, but some days later, when it has already started to fade.

Actually, the great range in magnitude is somewhat deceptive, because it applies only to visible light. Cool stars, such as Mira, radiate chiefly in the infra-red, and infra-red radiation does not affect our eyes. The variation at infra-red wavelengths is much less than in the visible range, and if we calculate all the wavelengths combined we find that Mira is only about two and a half times more powerful at maximum than at minimum. The luminosity ranges between 250 and over 1000 times that of the Sun, which is not very much when compared with a supergiant such as Betelgeux. The companion star is much feebler, but is also denser, and may be more massive than Mira itself even though it could be no larger than a planet such as Jupiter.

Mira's own mass is probably no more than twice that of the Sun. This means that its overall density must be very low indeed: about 0.0000002 that of the Sun, corresponding to what we normally call a laboratory vacuum. Of course, the core must be much more substantial, and this brings us on to the way in which Mira is evolving – and where it lies on the HR diagram.

Vast red stars were once regarded as young, newly-condensed out of the interstellar material and on their way to join the Main Sequence. This is incorrect, though we cannot blame the astronomers of a few decades ago for making the wrong choice. Mira, like Betelgeux, is well advanced in its life-story, but the two are not behaving in the same way. Betelgeux is changing in size and output only slightly, and over no very definite cycle. Mira, on the other hand, is pulsating. It is swelling and shrinking in a fairly definite and well-marked pattern, changing its surface temperature as it does so. It has left the Main Sequence, so that it has used up its main reserves of hydrogen 'fuel'. Different kinds of reactions near its core have either started, or are about to start.

There is also the point that a star of this kind will have to lose mass if it is to avoid a spectacular fate. Provided that we are right in assuming that some sort of shock-wave moves through the outer layers of the star after having originated deep inside the globe, there will be times when material is thrown off into space – remembering that in so vast and rarefied a globe, the outermost layers must be decidedly unstable. If the mass remaining when the star has used up all its nuclear reserves is more than about 1.4 times that of the Sun, the result will be a stellar collapse followed by a supernova outburst; if the mass left at the end of the giant stage is less than this, the star will merely shrink to become a white dwarf, as the companion of Sirius has already done and as the Sun will do in the future. There is strong evidence that red giant stars are losing mass quite rapidly, and Mira can be no exception.

Mira variables can be distinguished not only because of their redness and their long-period fluctuations, but also because of their spectra, which show changes in tune with the light cycle. They show some bright emission lines (due mainly to hydrogen) as well as the usual dark absorption lines. There is also the strong possibility that at times some of the light coming from the star is blocked out by obscuring material in its outer atmosphere.

Apart from Mira, the most interesting long-period variable is Chi Cygni, in the Swan (Fig. 42). It may be found between Sadr or Gamma Cygni, the middle star of the Cross of Cygnus, and the

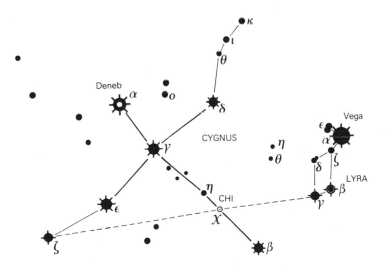

Fig. 42 Position of Chi Cygni

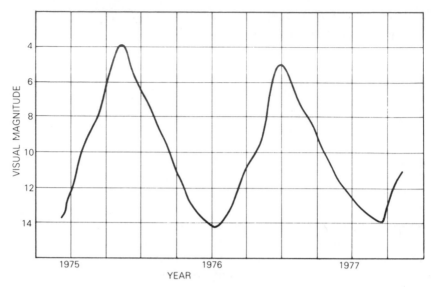

Fig. 43 Light-curve of Chi Cygni

lovely coloured double Albireo. Chi lies close to a normal fourth-magnitude star, Eta Cygni. It has a period of 407 days, much longer than Mira's, and it also has a greater range. At its peak it has been said to exceed the third magnitude, though I have never seen it do so, and am not confident that it ever rises above 4 (Fig. 43). At minimum it drops to below magnitude 14, so that it is lost not only to the naked eye and binoculars but also to small telescopes – particularly as it is situated in a rich field, and at its dimmest is not at all easy to pick out from the numerous faint stars in its neighbourhood. It is redder and cooler than Mira, and seems to be farther away and more luminous. It is extremely powerful in the infra-red range; if our eyes were sensitive to these long wave-lengths, Chi Cygni would be one of the brightest objects in the entire sky.

The main lesson we have learned from Mira stars is that like the less obviously red variables, they are stellar old-age pensioners rather than infants or even adolescents. We cannot watch a star evolving, unless it is undergoing some tremendous outburst; all we can do is to study stars at different stages in their careers, and then try to make sense of the overall picture.

Next time you have the chance, I suggest that you go out and take a careful look at the region near the Whale's head. Probably you will not see Mira unless you have optical aid and a star-map, but you may be lucky – and sooner or later you are bound to be rewarded with a good view of 'the Wonderful Star'.

X DELTA CEPHEI: STANDARD CANDLE IN SPACE

We have already met some of the characters of the Perseus legend. Strangely enough, one of the most obscure of them in the sky itself is King Cepheus, Andromeda's father. The stars marking the constellation lie between Deneb on the one side and Polaris on the other, but they are not bright – the leader, Alderamin, is only of magnitude 2.4 – and neither is there any really distinctive pattern. However, there are two objects of exceptional interest (Fig. 44). One is the red irregular variable, Mu Cephei, which has been nicknamed 'the Garnet Star'; binoculars or a telescope show that it fully justifies this description of it. It is of the same basic type as Betelgeux, but is much more luminous. One estimate gives its mass as twenty times that of the Sun, and it may be 100,000 times as luminous, while its globe would contain the whole orbit of Jupiter. It appears relatively faint only because it is so far away. But of even greater significance to astronomers is Delta Cephei, which also is variable, and whose changes in light were discovered by the remarkable John Goodricke in 1784.

You can find Delta Cephei easily enough. It is situated between Deneb and the central star of the W in Cassiopeia; it is one of a triangle of stars, the others being Zeta Cephei (magnitude 3.3) and Epsilon (4.2). Delta has a period of 5 days 8 hours 48 minutes, during which it ranges between magnitudes 3.5 and 4.4, so that at its best it is not much inferior to Zeta, the most prominent member of the triangle; at minimum it becomes fainter than Epsilon. Unlike Mira, it is absolutely regular. We know how it behaves, and we can always predict its brightness for any particular moment.

The light-curve is not symmetrical (Fig. 45). From minimum it takes about 1½ days to rise to maximum, while the fall back to minimum occupies about four days. Neither is the star red; the spectral type ranges between F and G, so at best it may be regarded

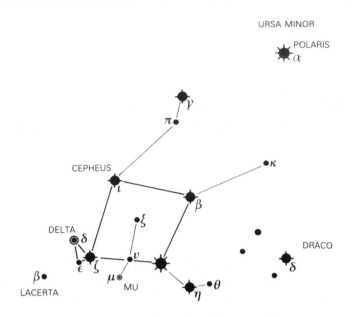

Fig. 44 Positions of Delta and Mu Cephei

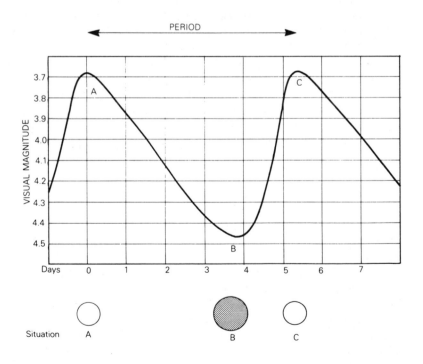

Fig. 45 Light-curve of Delta Cephei

as slightly yellowish. I admit that I have never been able to see any colour in it, either with the naked eye or with a telescope.

Shortly after Goodricke had made his announcement, the English astronomer, Pigott, found that another naked-eye star, Eta Aquilæ in the Eagle, fluctuates in the same way. Here the range is between 3.7 and 4.5, so that, like Delta Cephei, it is always an easy naked-eye object, and it too is completely regular, with a period of just over 7 days. It is particularly easy to identify (Fig. 46), because it lies not far from Altair, and directly between two slightly brighter stars, Theta Aquilæ (3.2) and Delta Aquilæ (3.4).

Many thousands of similar-type stars are now known, and they are always known as Cepheids, though if Pigott had made his discovery a few months earlier they might well have become known as Aquilids instead. Other naked-eye examples are Zeta Geminorum, in the Twins (maximum 3.7) and the southern Beta Doradûs in the Swordfish (4.5). Another southern short-period variable is Kappa Pavonis, in the Peacock. It can reach the fourth magnitude, though it is not of exactly the same type as Delta Cephei itself. There are many Cepheids within the range of binoculars or small telescopes, and because their light-changes are quick (at least in comparison with Mira or Betelgeux) they are fascinating to follow from night to night.

Cepheids are pulsating stars, but they are very different from the Mira variables. For one thing the average Cepheid is more luminous. At its peak Delta Cepheid could match 3000 Suns, perhaps more; its distance from us has been given as 1340 light-years. The spectrum, as we have seen, changes in harmony with the magnitude. When we plot Cepheids on the HR diagram, we find that they fall in a region between the Main Sequence and the red

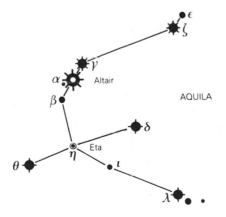

Fig. 46 Position of Eta Aquilæ

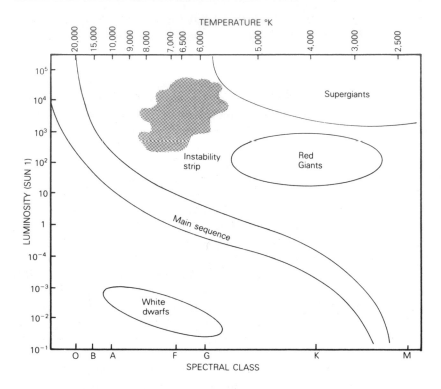

Fig. 47 The Instability Strip on the HR diagram

giants. This is called the 'Instability strip', which is a good term, because all the stars in it are variable (Fig. 47). In addition to the Cepheids (of which there are two varieties) there are the fainter short-period stars known as RR Lyræ variables.

There is direct observational proof that the Cepheids are alternately expanding and contracting, as was shown by the Russian astronomer A. Belopolski almost a hundred years ago. If the surface is expanding, it will be moving toward us, and the lines in its spectrum will be blue-shifted; when the surface is contracting, and therefore receding from us, the shift will be to the red, and the overall result is that the lines will oscillate around their mean position. Belopolski not only showed this, but also found that the period of oscillation was exactly the same as Delta Cephei's light period: 5.4 days. The temperature changes also followed the same cycle.

But why should a Cepheid behave like this? Clearly it has used up its main store of hydrogen fuel, and has left the Main Sequence. At this stage the core shrinks, under the influence of gravity, and heats up, while the outer layers expand and cool, changing the star into a red giant. At a core temperature of about 100,000,000°C, the helium which has been built up from the original hydrogen begins to react, using its nuclei to produce heavier elements. The whole process is decidedly complex, and the fate of the star depends upon how massive it is. With a comparatively massive star, the helium will start reacting very rapidly in what is called the 'helium flash', while with a star of lesser mass the onset of helium reaction will be slower. In either case the star, now a red giant, begins shedding its material.

When the star enters the 'instability strip' of the HR diagram, pulsations begin. The whole effect has been likened to that of a bouncing ball, but the ball will soon stop, whereas the Cepheids continue to swell and shrink over long periods of time. When the star is at its largest, the surface is at its coolest, and we have to reckon with the disrupted parts of atoms. If an atomic nucleus loses some or all of its circling electrons it is said to be ionized, and is termed an ion; when the Cepheid is coolest, the ions combine with spare electrons to build up complete atoms again. Complete atoms are electrically neutral, and these neutral gases radiate energy into space, so that the internal pressure drops and the star begins to shrink. But this cannot go on for long. As the gases inside the star become denser, they become ionized once more, and we are back to the start of the cycle. The whole process will go on until the supply of the new 'fuel', helium, begins to run low.

It does not take much intuition to see that the behaviour of a Cepheid depends not only upon its mass but upon its luminosity. This has led to one of the most important of all discoveries in modern astrophysics. The discovery itself was made more or less by accident, though this does not in any way detract from the achievement of the astronomer responsible, Henrietta Swan Leavitt.

Miss Leavitt was American; she was born in Lancaster, Massachusetts, in 1868, and took her degree at the age of twenty-four. Her first work in astronomy was voluntary. She went to the Harvard College Observatory, and worked with Edward Pickering upon stellar spectra and cataloguing. In 1902 she was made a permanent member of the staff, and worked there until, sadly, she died of cancer in 1921. During her career she discovered over 2400 variable stars and four novæ, and became known as an expert theorist as well as an observer.

She was very interested in the two galaxies which we know as the Magellanic Clouds, and which are the nearest of all the external systems; they have been regarded – whether justifiably or not – as satellite systems of our own Galaxy. With the naked eye they are very prominent, and the larger of the two can be seen even in moonlight. Superficially they look rather like broken-off parts of the Milky Way, and they are familiar to all those who live in the southern hemisphere; unfortunately for Europeans and North Americans, they lie not far from the south celestial pole – the Large Cloud mainly in Dorado and the Small Cloud in Tucana, the Toucan.

Obviously, then, they are permanently invisible from Harvard, but before the First World War an outstation of the observatory had been set up at Arequipa, in Peru, and a photographic survey of the sky had been carried out from there. The telescope used was of modest size; it was a 10-inch refractor, known as the Metcalf Telescope, which is still being regularly used, though it has long since been transferred to the observatory of Boyden, near Bloemfontein in South Africa. It is in excellent order, and I made some observations with it myself a year or two ago.*

Around 1908 Miss Leavitt obtained large numbers of photographs from Arequipa, taken with the Metcalf refractor, and decided to make a very careful study of the Small Magellanic Cloud. She was particularly concerned with variable stars; would she find any in the Cloud? She did – and many of them were Cepheids. Because she had so many photographs, taken at different times, she was able to find out how the Cepheids were fluctuating, and to determine their periods. Gradually it dawned upon her that she was obtaining some remarkable results. The Cepheids which had the longer periods were always brighter than those with shorter periods, and there seemed to be a well-marked relationship.

At first there was the chance that the correlation was due to nothing more than coincidence. After all, the periods of Cepheids show a wide range, from a few days up to several months, and they differ in brightness; but with the Small Cloud, it could be assumed that all the variables were at the same distance from us. Of course this is not quite true, but it was good enough for most purposes, just as in the ordinary way it is enough to say that Bristol and Bath are the same distance from New York. The distance of the Small Cloud had not then been measured, and neither was it known whether the Cloud belonged to our Galaxy or was some way beyond, but there could be no doubt that it was much too remote to show any detectable parallax.

Finally, in 1912, Miss Leavitt published a paper in which she gave details of twenty-five Cepheids in the Small Cloud. Their average period was around 5 days, but the full range was from 1.2 days up to 127 days, and there could be no doubt that the link between

---

*The adjoining dome houses a slightly larger telescope, a 13-inch refractor, and I was informed that under the floorboards there lives a 4-foot cobra. I am delighted to say that I never saw it, but I heard the story of how a visiting astronomer, entering the dome ahead of his escort, was confronted by a rearing snake. I gather that his observing programme that night was somewhat curtailed!

period and apparent magnitude – and, hence, period and luminosity – was real. Ejnar Hertzsprung and Harlow Shapley then converted Miss Leavitt's results into a table which would give the real power of a Cepheid as soon as its period had been found. There was no reason to suppose that the Cepheids in the Small Cloud were different from Cepheids anywhere else, so that astronomers were at once provided with what might be termed 'standard candles' in space.

The immediate drawback was that no Cepheid is close enough to show measurable parallax, so that everything depended upon studies of their spectra. Moreover, there is considerable absorption of light in space due to interstellar matter (a factor which was badly underestimated in the early days), and this of course affects the brightness of a Cepheid or any other star. But the final results were immensely significant. By 1918 Shapley was able to show that the Small Cloud was about 94,000 light-years away. This is too little – the real distance is 190,000 light-years – but it was much greater than astronomers had expected. By then, however, Shapley was busy using the Cepheids for another equally important investigation: measuring the size of the Galaxy itself.

William Herschel, discoverer of the planet Uranus in 1781, had been the first to give a reasonable picture of the shape of the Galaxy, which he likened to a 'cloven grindstone'. When we look along the main thickness of the system, we see many stars in almost the same line of sight, which gives the appearance of the Milky Way band. But as well as containing about a hundred thousand million stars, the Galaxy also includes open and globular clusters, nebulæ and many other kinds of objects. Open clusters, such as the Pleiades, may be fairly local, but the globular clusters are not. They are symmetrical, and are composed of stars which look as if they are so closely crowded together near the centre of the system that they are in serious danger of colliding. This is not so – the stars even in the most densely-populated regions are still light-months or at least light-weeks apart – but it is true that if we lived on a planet orbiting a star inside a globular cluster, the night skies would be glorious. There would be no darkness, because of the presence of many stars brilliant enough to cast shadows. Also, many of these stars would be red, because in an old system such as a globular cluster the leading members have already left the Main Sequence and moved into the giant branch of the HR diagram.

Shapley realized that the globulars are not uniformly distributed across the sky. They are commonest in the southern hemisphere, and the two brightest examples are permanently out of view from European or North American latitudes. One is Omega Centauri, which is not far from the Southern Cross and is an easy naked-eye object; the other is 47 Tucanæ, which appears almost superimposed upon the Small Cloud of Magellan. In the northern hemisphere the only naked-eye globular is Messier 13, in Hercules. The main concentration of globulars is in the direction of the star-clouds in Sagittarius, the Archer.

Shapley was an expert observer. He had been born in Missouri in 1885, and had become an astronomer by accident! He had intended to make a career out of journalism, but the school in which he wanted to enrol had no vacancy for another twelve months, and Shapley took up astronomy to mark time. He never changed back. In 1914 he went to the Mount Wilson Observatory in California, where he was unable to use the giant telescopes – the 60-inch reflector and, from its completion in 1917, the 100-inch Hooker reflector which was for so many years in a class of its own. The 100-inch has had a glorious career; we can only hope that its present withdrawal from active service, due to the increased light pollution from nearby Los Angeles, is no more than temporary.

Shapley hoped to measure the size of the Galaxy by finding out the distances of the globular clusters which lay around its edge. Therefore he began a careful search for Cepheids inside the clusters, and he found them. It was a laborious task, needing a tremendous amount of care and patience, but Shapley was not to be denied, and at last he was able to give the first realistic estimate of how large the Galaxy really is. It was, he said, about 300,000 light-years from one side to the other. The centre of the system lay beyond the lovely star-clouds in Sagittarius, and this, of course, is where we see the greatest numbers of globular clusters.

Though Shapley overestimated the size of the Galaxy, inasmuch as the real overall diameter is only 100,000 light-years (perhaps slightly less), his results were of the right order, and it effectively disposed of the previously held theory that the Sun lay near the centre of a system no more than 10,000 light-years across. It was by no means his only contribution to astronomy; there were many others, and his early passion for journalism showed itself in his popular books and articles. He was also an excellent speaker. I well remember that not long before he died, in 1972, he joined me in one of my *Sky at Night* television programmes, and described how he had managed to measure the size of the Galaxy more than half a century before.

There was, however, the problem of the so-called 'starry nebulæ', and here Shapley was initially wrong.

Nebulæ are of two kinds. There are the gaseous clouds, such as the Sword of Orion, which, as we have seen, are stellar nurseries. But other objects in Messier's original catalogue were quite different. They do not look like gas clouds; they seemed to be made up of stars. Such was the famous 'Nebula' in Andromeda, Messier's 31st object, which you can just see with the naked eye on a clear night and which is very distinct with binoculars, though photographs taken with large telescopes are needed to show it really well. William Herschel made the inspired guess that these starry nebulæ might be independent galaxies, far beyond the Milky Way, but at that time there was no way of proving it, because there seemed to be no chance of measuring their distances.

Next in the story came a remarkable man, the third Earl of Rosse. He was an Irish aristocrat, whose family seat was at Birr, a few miles from Athlone in what is now Eire. It was his determination to make what would be the largest telescope in the world, and he succeeded: in 1845 he completed a monster reflector with a metal mirror 72 inches across. It was an extraordinary instrument, mounted between two massive stone walls, so that it could swing for only a limited distance to either side of the north-south line; it was awkward to use, and there was no way of driving it satisfactorily, so that everything had to be hand operated. But it worked well, and Lord Rosse achieved everything which he had hoped. It was also said that nobody who came to him for help or advice ever went away unsatisfied.

Lord Rosse swung his giant telescope toward some of the starry nebulæ, and found, to his surprise, that many of them were spiral in shape, like Catherine-wheels. His drawings were remarkably accurate, and when we compare them with modern photographs we can appreciate how good an observer he was. The Andromeda system is itself a spiral, though unfortunately it lies more or less edge-on to us, and its full beauty is lost. If it were face-on, at the same angle as the 'Whirlpool' M51 in the little constellation of the Hunting Dogs (close to Alkaid in the Plough), it would be really splendid.

For decades there was no telescope in the world, apart from the Birr reflector, which could show the spirals, but later in the nineteenth century came modern-type refractors, and in our own century these were in turn succeeded by powerful reflectors, of which the Mount Wilson 100-inch was the most impressive. These could resolve the starry nebulæ, but still their distances could not be found, and Shapley for one was under the impression that they were comparatively minor features, contained in our own Milky Way system.

Parallax methods were useless; the spectroscope was at first of limited help, and the only way to clear up the problem was by using those invaluable standard candles, the Cepheids. This was the task which Edwin Powell Hubble set himself.

Hubble, a native of Missouri, was born in 1889, and for a while practised law (incidentally, he was also an excellent amateur boxer). He then went to Yerkes, but after the First World War, during which he served in the US infantry, he took up a post at Mount Wilson, and remained there for the rest of his career.

Hubble was fascinated by the starry nebulæ, and also by Henrietta Leavitt's work on the Cepheids. If he could only find these useful stars inside the spirals, he reasoned, he would know the answer – but this was easier said than done. Powerful though the Cepheids are, they were hard to distinguish in remote systems, and Hubble's only chance lay with the 100-inch reflector, with its tremendous light-grasp. He persevered, and in 1923 he detected the first variable star in the Andromeda Spiral; shortly afterwards he found more, of which a dozen were Cepheids. As soon as he had measured the periods of these Cepheids, he realized that they were much too remote to be members of our Galaxy. He estimated their distances – and, hence, the distance of the Spiral itself – at 900,000 light-years. We now know that this was too small, but it was enough to show that the systems really were independent galaxies; so were the other 'starry nebulæ', many of which were not spiral in form.

It would be impossible to overestimate the importance of this discovery, probably the most significant made since Kepler had conclusively proved that the planets move round the Sun and not round the Earth. And Hubble had only begun his work; it was he, too, who showed that the entire universe is expanding – a point to which I will return later. He was still working hard at observation and theory almost up to the time of his death in 1953. By then there had been another significant development, which was quite unexpected. It was due to a Mount Wilson colleague of Hubble's, Walter Baade, who was German by birth but who spent much of his life in the United States.

Baade, like Hubble, was intrigued by the problem of the galaxies (the old term of 'spiral nebula' had been dropped; a galaxy is not a nebula), and he had the uneasy feeling that something was wrong. Assuming that the Andromeda Spiral lay at a distance of around 900,000 light-years, it would be much smaller than our Galaxy, and so would all the other external systems. Remember that every time Man has claimed a position of special privilege, he has been shown to be humiliatingly wrong. The Earth is not important; neither is the Sun, so why should our Galaxy be larger than the rest? It did not seem to be logical.

Moreover, there were other short-period variables in the instability strip of the HR diagram which added to the doubts. These were the RR Lyræ stars, named after the first-discovered member of the class, a faint star in the constellation of Lyra. Their periods were short, from half a day to a day and a quarter, and all of them seemed to be of the same luminosity, about a hundred times that of the Sun. They also could be used as standard candles, but there were none to be found in the Andromeda Spiral or any of the other galaxies. Why not?

With the Mount Wilson telescope, Hubble had studied the bright stars in the outer parts of the Andromeda Galaxy, most of which were hot and blue, and also the bright patch in the centre of the system. What he had not been able to do was to distinguish individual stars in the central regions; they merged together in a general blur. Baade determined to try. In 1943, during the Second World War, Los Angeles was blacked out, and the skies over Mount Wilson were exceptionally dark, so that Baade was able to make use of the full power of the 100-inch. (Though German-born, he was not restricted in his astronomical activities.) He took every possible precaution; even so he only just 'got under the fence', as he put it, but eventually he was able to show that the stars near the centre of the Andromeda Spiral were mainly old red giants and supergiants. In fact there were two kinds of 'stellar populations'. In Population I areas the leading stars were hot and bluish-white; there was a good deal of interstellar material, and star formation was still going on. In Population II areas the leaders were red stars which had left the Main Sequence, and all the star-forming material had been used up. Of course the boundaries between the two types of region were not sharp, but generally speaking the outer parts and the spiral arms of galaxies were Population I, while the centres and the globular clusters were Population II.

After the war, Baade was able to use the new 200-inch reflector on Palomar Mountain, and established that there had been a serious error. There were Population I Cepheids and Population II Cepheids, but the period-luminosity relationship was not the same for both. The Population II stars were far less powerful than their Population I counterparts.

Now you can see the complication. The distances of the galaxies had been calculated on the assumption that the Cepheids in them were of Population II. In fact they were of Population I, so that they were much more luminous, and much more remote, than had been believed. Baade announced his results in 1952, at a meeting of the Royal Astronomical Society; in a short paper he calmly doubled the size of the universe. I can still remember the stunned silence as he finished speaking and sat down!

The modern value for the distance of the Andromeda Galaxy is 2.2 million light-years. Even so, it is one of the very closest of the external systems, and is a member of what we call the Local Group. It is also considerably larger than our Galaxy, with more than our quota of 100,000 million stars.

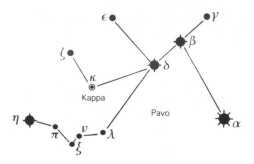

Fig. 48 Position of Kappa Pavonis

Population II Cepheids (otherwise known as W Virginis stars) are common enough; there is one naked-eye example, Kappa Pavonis, which has a period of just over 9 days (Fig. 48). This might be expected to make it more luminous than Delta Cephei (5.3 days) or Eta Aquilæ (7.1 days), but in fact Kappa Pavonis is 'only' about three times as powerful as the Sun, and is a true glowworm compared with a Population I Cepheid of the same period.

The brighter Cepheids can be detected out to great distances, and they can be used to examine galaxies out to well over 10 million light-years. It is only with the more remote systems that they are lost in the general background blur, and we have to use less direct, less trustworthy methods of distance gauging. But in any case, it was the Cepheids which set us on the right track, thanks to astronomers such as Henrietta Leavitt, Hertzsprung, Shapley, Hubble and Baade. We would indeed be at a loss if we did not have the invaluable 'standard candles' to guide us.

XI CN TAURI: BIRTH OF THE CRAB

Of all the sky-watchers of ancient times, there were few who were more diligent than the Chinese. Their records go back a very long way indeed, and their careful notes about phenomena such as eclipses and comets are of tremendous value. (For example, they are our sole authorities for descriptions of Halley's Comet before the return of BC 240, and they may even have seen it as early as BC 1059.) Subsequently, Chinese science declined, and in mediæval and near-modern times it has been very much in the background – quite apart from Chairman Mao's cultural revolution, which consigned all scientists to the scrap-heap. Nowadays the situation is changing fast. Chinese astronomy has come back into the picture, and there is even talk of China launching artificial satellites which cannot be put into orbit by the accident-prone Shuttle or Europe's Ariane.

But not all Chinese records were stopped after the great days of the Empire were over. In the summer of 1054 the Court Astrologers in Pekin (now called Beijing) made an unexpected discovery. There, in the constellation we call Taurus (the Bull) was a new star, shining so brilliantly that it could not be overlooked, and which had most certainly not been there before. A report to the Emperor, made by the astrologer Yang Wei-te on a date corresponding in our calendar to 27 August, runs as follows:

> Prostrating myself before Your Majesty, I hereby report that a guest star has appeared; above the star in question there is a faint glow, yellow in colour. If one carefully examines the prognostications concerning the Emperor, the interpretation is as follows. . . . Its brightness means that there is a person of great wisdom and virtue in the country.

Evidently Yang Wei-te would have had nothing to learn from Modern Civil Servants! Whether his report is wholly reliable must be

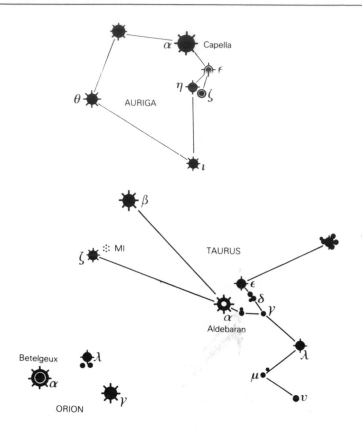

Fig. 49 Position of the Crab Nebula, M1

considered doubtful (for example, yellow was the official colour of the ruling Sung Dynasty, and so obviously it was a good idea to describe the 'guest star' as yellow). There had been similar outbursts before, in the years 185, 285, and 1006; the last of these, in the constellation of Lupus (the Wolf) had become so brilliant that apparently it rivalled the quarter-moon, and remained visible for two years, though unfortunately it was so far south in the sky that it was poorly documented. The position of the 1054 star, on the other hand, was given reliably, close to the third-magnitude star Alheka or Zeta Tauri. Apparently it lasted for 653 nights before dropping below naked-eye visibility, and for 23 days it was visible even when the Sun was above the horizon, so that it equalled or more probably surpassed Venus at its best. The modern designation for it is CN Tauri.

We do not depend entirely upon the Chinese. There is also a record of the 'guest star' from Japan, and another from Arabia, while it has also been suggested that it may be shown on some Red

Indian cave-drawings found in the southern United States. Of course, it was lost as soon as it faded below the sixth magnitude; telescopes were not to come to the rescue for more than five hundred years in the future.

The Chinese recorded another 'guest star', in Cassiopeia, in 1181; there was a violent outburst in the same constellation in 1572, about which we know a great deal because it was fully described by the Danish astronomer, Tycho Brahe; and yet a third appeared in 1604, to be observed by many people including Johannes Kepler. All these stars were exceptional, and clearly different from the normal run of what we call novæ. Yet by sheer bad luck, all of them burst forth in pre-telescopic times, and it was by no means certain what they really were.

Then, in 1731, an English doctor named John Bevis, who was an enthusiastic amateur astronomer and had built his own observatory, made a discovery which was far more important than he could have guessed. Using one of the small-aperture, long-focus refractors which were common at the time, he looked in the region near Zeta Tauri and found a dim, glowing patch of light. He was working on an atlas of the sky, but its publication was delayed because the printers had financial problems, and it did not finally appear until 1786, after Bevis' death. In the meantime, Bevis had written to the French astronomer, Charles Messier, telling him about the strange patch. Messier may already have discovered it independently; at any rate, he described it as 'nebulosity above the southern horn of Taurus. It contains no star; it is a whitish light, elongated like the flame of a taper, discovered while observing the comet of 1758. Observed by Dr. Bevis in about 1731'. In Messier's great catalogue of clusters and nebulæ, the patch was given pride of place; we know it as Messier 1, or simply M1.

Once attention had been drawn to it, astronomers everywhere saw it, and in fact it is not in the least difficult. Most books say that it is invisible without the use of telescope, but I can see it clearly with good binoculars, somewhat north-west of Zeta Tauri and in the same field of view. It was, of course, drawn by Lord Rosse in the 1840s, and it was he who commented that its shape reminded him of a crab, so that today everyone calls it the Crab Nebula.

Lord Rosse believed that with sufficient magnification the Crab could be resolved into stars, but for once he was wrong, and the great twentieth-century telescopes showed nothing more than a filmy mass. It was first photographed in 1892, and in 1913 Vesto Melvin Slipher, director of the Lowell Observatory in Arizona, studied its spectrum, proving that the Crab is made up of gas rather than stars. He also believed it to be associated with a strong magnetic field. Even when Edwin Hubble had shown that the 'starry nebulæ' are external galaxies, the Crab remained a puzzle. Certainly it was not a galaxy, but neither did it seem to be an ordinary gaseous nebula, such as the Sword of Orion or the vast Tarantula Nebula in the far south of the sky.

C.O. Lampland at Lowell, and J.C. Duncan at Mount Wilson made further studies, and discovered that there were significant changes taking place in the Crab over periods of only a few years. The gas cloud seemed to be expanding; and by working backwards, so to speak, it began to look as though the expansion had begun about 900 years earlier. The Swedish astronomer, Knut Lundmark – one of the great pioneers in studies of our Galaxy and others; he is not nearly so well-known today as he ought to be – made the casual comment that the position of the Crab agreed well with that of the 'guest star' of 1054. Edwin Hubble repeated the suggestion in 1928, but it was not until modern times, with the arrival of radio astronomy, that all doubts vanished. Today we are absolutely sure that M1 is indeed the remnant of the Chinese star; it is all that is left of what is called a supernova outburst.

The science of radio astronomy began in 1931, when Karl Jansky, an American radio engineer of Czech descent who was working for the Bell Telephone Company, was studying 'static' – the wireless operator's enemy – by means of a weird aerial which he had nicknamed the Merry-go-Round, and which he had built partly out of the parts of an old Ford car. To his surprise he found that there was a steady, continuous hiss which did not seem to be of terrestrial origin. Before long he established that it came from the Milky Way – to be more precise, the region of the star-clouds in Sagittarius, which indicate the direction of the centre of the Galaxy. Jansky published some papers about it, which aroused about as much overall scientific interest as a feather falling on to damp blotting-paper, and after the mid-1930s he turned his attention elsewhere. It was not until after the end of the war that interest in long-wavelength astronomy was really kindled, mainly because English research workers led by J.S. Hey had found 'radio noise' which they at first attributed to German jamming of our radar, but which they then decided must come from the Sun.

One point is worth making here. We read a great deal about radio 'noise', and it has been broadcast often enough, but the actual hissing sound is produced in the apparatus itself. Sound cannot travel through a vacuum, and we cannot receive any whisper of noise from beyond the Earth. What happens is that the long-wavelength radiations are collected and then analysed in various ways, one of which (not the best) being to convert the signal to sound. Most of the true analysis is done by means of a graph traced out by a pen upon a moving paper drum. The name 'radio telescope' is also somewhat misleading, if only because the instrument does not produce a visible picture of the object being studied, and one cannot look through it. Radio telescopes are of many different designs, but all of them have the same basic purpose.

An early problem was that apart from the Sun, discrete radio sources in the sky did not seem to be linked with visible objects. Brilliant stars such as Sirius and Vega remained obstinately radio-quiet; the emissions seemed to come from regions unmarked by anything in particular. But as radio methods improved, and resolution became sharper, identifications started to be made. One of the earliest, in 1948, was that of the Crab Nebula, which has turned out to be one of the strongest sources in the sky.

Both the visible spectrum and the radio emissions indicated that there was an unusual kind of power-source involved. The Russian astronomer, Iosif Shklovskii, suggested that this might be due to electrons moving in a powerful magnetic field, and we now know that he was correct; the official term for it is 'synchrotron radiation'. And as the years went by, it became more and more evident that the Crab was sending out radiations at almost all wavelengths. X-rays were detected in April 1963, from the instruments carried on board a rocket which soared briefly above the Earth's atmospheric screen; there were also infra-red radiations, ultra-violet, and even the extremely short, highly energetic gamma-rays.

Altogether the Crab seemed to be in many ways unique. The distance was estimated as 6000 light-years, which meant that the Chinese supernova actually happened about the year BC 4000, before there were any written records; but what had become of the star itself? Would there be any trace of it deep inside the expanding gas-cloud?

The first problem to attack was that of the nature of the explosion itself. There have been many cases of what we call novæ – a term which really means 'new', but is inappropriate because a nova is not genuinely a new star at all. It consists (Fig. 50) of a binary system, one member of which is an ordinary Main Sequence star while the companion is a highly evolved white dwarf. The white dwarf steadily pulls material away from its larger, less dense companion, and this material builds up around the dwarf until the whole situation becomes unstable; nuclear reactions are triggered off, and the white dwarf suffers a sudden outburst, which makes it flare up to many times its normal brilliance before fading slowly back to obscurity. The last really bright nova was that of 1976, in Cygnus, which blazed up from extreme faintness to above the second magnitude in only a few hours. I was an independent discoverer of it, but I cannot claim priority, as I must have been about eightieth on

111

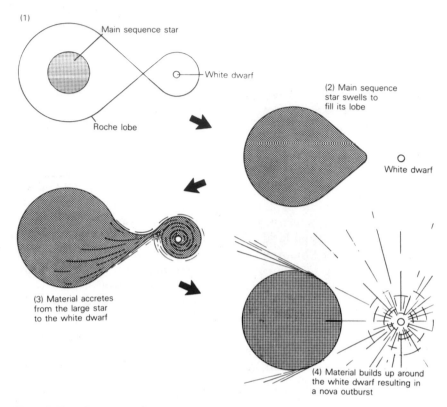

(1)

Main sequence star

White dwarf

(2) Main sequence star swells to fill its lobe

Roche lobe

O
White dwarf

(3) Material accretes from the large star to the white dwarf

(4) Material builds up around the white dwarf resulting in a nova outburst

Fig. 50 Development of a nova

the list. The star was first seen on the afternoon of 29 August by Japanese observers, before darkness fell over Europe; as soon as the sky over England became dark, almost every serious amateur astronomer noticed it – as well as a few professionals. It did not last for long. Within a few days it had faded below naked-eye visibility, and it has now become very dim indeed.

There have been various other novæ during the past decades. In 1901 Nova Persei attained the first magnitude; in 1918 Nova Aquilæ, in the Eagle, became brighter than any star in the sky apart from Sirius; in 1934 a new star in Hercules outshone Polaris for some weeks, and so on. Telescopic novæ are not infrequent (and it must be added that amateur observers have a fine record in discovering them). But the 1054 outburst, and those of 1006, 1572 and 1604, were quite different. They were not normal novæ; they were supernovæ.

A supernova is not only much more violent; but involves the virtual or complete destruction of a star. There are two types. In Type I, we are back to the normal-star-plus-white dwarf combination, but this time the build-up of material accumulated by the white dwarf is so great that it triggers off an outburst in the body of the star itself. There is an explosion which literally blows the white dwarf to pieces. For a brief period the output of energy may equal 18 million Suns, and it takes a long time for the fading to be complete. I will have more to say about this in Chapter 15.

With Type II supernovæ, we have only one star, but it is much more massive than the Sun, so that it runs through its life-cycle much more quickly. After the Main Sequence stage, when the star becomes a red giant, it radiates by building up heavier and heavier elements inside it, until at last the main core is made up of iron. This is where the real crisis begins, because iron cannot be forced to release energy no matter how high the temperature and how great the density. The core does its best to support itself against the tremendous pressure of the overlying layers, but in the end it fails, and it collapses when the density reaches about a thousand million grammes per cubic centimetre. In less than a second the density soars still further, and the various parts of the broken-up atoms are not only packed together (as in a white dwarf) but experience an even stranger fate. The protons and electrons are squeezed together. Since a proton has a positive charge of electricity and an electron has a negative charge, the result is a neutron with no charge at all ($+ 1 - 1 = 0$). Matter as dense as this cannot be further compressed, but the outer portions of the star are still collapsing at an appreciable fraction of the velocity of light. When they crash into the unyielding core, there is a violent shock-wave; when the wave reaches the star's outer layers, these layers are thrown off, and the output of energy is almost as great as with a Type I supernovæ. The end product is a cloud of expanding gas, which will eventually dissipate and be lost, plus the remnant of the old star, now made up of neutrons. It does not need much imagination to realize that this is an exact description of the Crab Nebula. We can see the gas-cloud; all we have to do now is to locate the neutron star inside it.

In fact, neutron stars had been known as theoretical possibilities for a long time, but their actual discovery in 1967 was more or less accidental. At the Cambridge radio astronomy observatory, a team led by Professor Antony Hewish had been making observations with a radio telescope which was not a dish, such as the famous instrument at Jodrell Bank, but was made up of what looked superficially like a collection of barbers' poles covering a wide area. In August, Jocelyn Bell, one of the team, suddenly noted a peculiar radio source which was detected quite regularly. It was switching on and off at an amazing speed – only 1.34 seconds – and it seemed to be so regular and so unusual that it was at first assumed to be artificial.

Jocelyn Bell persisted, and – with difficulty – interested other members of the team. The strange source was nicknamed a pulsar, because it was pulsating so rapidly. It could not be due to anything on or near the Earth; could it be a signal from intelligent beings far across the Galaxy? Briefly, this idea was seriously considered, and became known as the LGM or Little Green Men Theory. It is also true that the Cambridge team delayed making any public announcement until they were quite sure that the source was natural.

Several more pulsars were soon found; their periods were all very short, and all seemed to be gradually slowing down, so that presumably they were losing energy. The shortening was tiny – a fraction of a second per century – but by now we have clocks which are better timekeepers than the Earth itself, and the increases in the periods of the pulsars could be measured, though now and then there were 'glitches' when a pulsar temporarily speeded up again.

Having disposed of the little green men, astronomers cast around for a better explanation. At Jodrell Bank, where the most famous of all radio telescopes had been set up, the favoured view was that a pulsar could be simply a white dwarf which was spinning round very rapidly, and was sending out radio emissions from only one or two points on its surface. This idea also had to be abandoned when theory showed that even though white dwarfs are very small, they are still much too large to spin as rapidly as the pulsars pulse; they would immediately break up. The obvious alternative was a neutron star, only a few miles in diameter. Then, in 1968, a pulsar was detected inside the Crab Nebula, with a period of only 0.33 of a second, the shortest known. This would fit in well with the idea that the pulsar was only born, in its present form, at the time of the Chinese supernova. At a mere 6000 light-years, it was presumably the youngest pulsar so far detected, so that it would have lost less of its original energy.

This was important, because there was at least a chance that the pulsar would be observable optically; it flashed at radio wavelengths, and might also flash in visible light. At the Steward Observatory in Arizona three astronomers, W.J. Cocke, M.J. Disney and D.J. Taylor, decided to hunt for it, using the Observatory's 36-inch telescope. For four nights they were unsuccessful, and their results were completely negative. As their allotted time was up, they dismantled their equipment and prepared to depart, feeling rather disconsolate.

Then, unexpectedly, news came through that the observer next on the observing rota had been delayed, because his wife had been taken ill. More in hope than in expectation, the team re-assembled their equipment and decided to make use of the extra two nights with which they had been presented. In checking, Disney suddenly realized that there had been an elementary mistake in setting up the equipment; the correction for the Earth's movement had been fed into the computer with the wrong mathematical sign. Once the error had been corrected, the team went back to work. They were temporarily thwarted by clouds; then, in the evening, the sky cleared – and in the first ten minutes of useful observing, the tiny flashing of the Crab pulsar was identified. It was by then early in the morning of 16 January 1969. Within a few days the identification had been confirmed at other observatories.

There was one note of frustration. Roderick Willstrop, at Cambridge, had made the same sort of scan of the Crab in November 1968, but had not been able to analyse his observations because of lack of available time on the university computer. When he did – after the American announcement – he found that the Crab pulsar was there, flashing quite perceptibly. The discovery had been within his grasp.

Since then a few more pulsars have been optically identified. There is one in the Gum Nebula, in the southern constellation of Vela, which is undoubtedly a supernova remnant, because the gas-patches are still evident even though the outburst took place in prehistoric times. (The nebula is named after the Australian astronomer, Colin Gum, who paid special attention to it; sadly, his career was cut short by his death in a skiing accident.) The Vela pulsar was first recorded in visible light in 1977 from the Siding Spring Observatory in New South Wales. It is dimmer than the Crab pulsar, and with a magnitude of 26 it is about the faintest object ever detected from Earth. The supernova itself must have been spectacular, and it is a pity from our point of view that it occurred about 11,000 years too early!

But what exactly is a neutron star, and why does it pulse? The accepted answer is that the star is rotating, and that it is sending out its radio emissions from its magnetic poles, which do not coincide with the poles of rotation. Every time the Earth passes through a beam, we receive a pulse, just as an observer on the sea-shore can receive regular flashes from the beam of a distant rotating lighthouse (Fig. 51).

The diameter of a neutron star is usually no more than a dozen miles or so (20 km); the density may be at least 100 million million times that of water; the outer layers are thought to be solid or even crystalline, while the material below is composed chiefly of neutrons, and the core could be made up of 'hyperons', about which our ignorance is virtually complete. The quick rotation can be accounted for by what is usually called the 'skater' analogy. If a skater whirls round, extending his arms, and then folds his arms beside his body, his rate of spin will increase. It is the same with a collapsing star; as it shrinks, its rotation speeds up, so that any

115

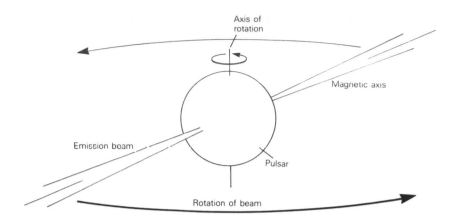

Fig. 51 Pulsar principle

material less dense than that of a neutron star would be unable to hold together.

So the Crab has told us a great deal. Its pulsar was the first to be detected optically, and we now know – or think we know – how it acts as the Crab's 'power-house'. But it will not last indefinitely. It is slowing down and losing energy, and eventually it will stop pulsating, to end its career as a cold, dead globe. By then the gas-cloud will have been lost in space, and there will be no visible trace left of the Crab Nebula.

But for the moment, and for the foreseeable future, it is there for our inspection. To study it properly you need the beautiful, detailed photographs taken with powerful telescopes, showing the incredibly complex structure which Lord Rosse likened to the shape of a crab; but even with a pair of binoculars you can find the dimly shining cloud, and it takes an effort of the imagination to realize that for a brief period, less than a thousand years ago, this was the brightest object in the whole sky apart from the Sun and the Moon.

XII ETA CARINÆ: ERRATIC STAR

At its peak, a supernova may shine as brilliantly as all the other stars in its galaxy put together. We can see them in external systems, but astronomers always bemoan the fact that there have been no observable supernovæ in our Galaxy since the invention of the telescope. They may have occurred, and there is evidence that the strong radio source which we call Cassiopeia A was produced by a supernova which we would have seen around the year 1670 but for the fact that there was too much interstellar dust in the way. Yet we have been denied the opportunity to study one of these titanic outbursts from close range – by which I mean a few thousands of light-years; anything much nearer would be too close for comfort!

By the law of averages we seem to be about due for a supernova. If they occur once in every 600 years, and the last of which we have definite proof was in 1604, the stage ought to be set. But averages mean nothing; we could well have two galactic supernovæ within a week and then no more for thousands of years. So is there any possible supernova candidate which we can identify? There is one, though we are admittedly working upon nothing more than intelligent speculation. Our potential supernova is Eta Carinæ, which already has the reputation of being the most erratic variable in the whole of the sky (Fig. 52).

Sadly, it is too far south to be seen from Europe. It lies in the constellation of the Keel, which is itself part of a vast constellation which was chopped up, by express order of the International Union in 1932, because it was too unwieldy. It was called Argo Navis, the Ship Argo, commemorating the vessel which took Jason and his companions on their rather unprincipled quest of the Golden Fleece. Argo covered a tremendous area; its leading star was Canopus, inferior in apparent brightness only to Sirius, and there were also many stars of the second and third magnitudes, plus a rich part of

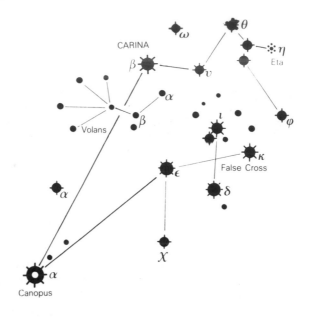

Fig. 52 Position of Eta Carinæ

the Milky Way. Finally it was dismantled into Carina (the Keel), Vela (the Sails) and Puppis (the Poop). The brightest section is Carina; Canopus, which used to be known as Alpha Argûs, has now become Alpha Carinæ. Here, too, we have the False Cross, sometimes mistaken for the Southern Cross, which lies partly in Carina and partly in Vela. Not far from it is Eta Argûs, now renamed Eta Carinæ.

It has a strange history. It was noted in 1677 by Edmond Halley, who had made a special trip to the island of St Helena to catalogue the southern stars which never rose over England. Halley rated Eta as of the fourth magnitude, but there seemed to be nothing exceptional about it. In 1751 the French astronomer, Lacaille, also in southern latitudes for star cataloguing purposes, ranked it as of magnitude 2. It then declined, and we know that from 1811 to 1815 it sank to the fourth magnitude before brightening up again. The real period of splendour began in 1827, when Burchell, in Brazil, found it almost equal to Acrux, the brightest star in the Southern Cross, so that it had attained the first magnitude. Following another slight decline, it blazed out in 1837. On 16 December of that year Sir John Herschel was at the Cape of Good Hope, where he had taken a large telescope so as to make the first really detailed and comprehensive survey of the far-southern stars (in which he was very successful). When he looked at Eta, he discovered that it was as bright as Alpha Centauri, so that only Sirius and Canopus surpassed it. Another fading, to about equality with Aldebaran in Taurus; then, in April 1843, the climax – from all accounts Eta Carinæ outmatched Canopus, and almost equalled Sirius. The magnitude must have been brighter than −1. Then, slowly, a decline set in. By 1860 Eta was barely visible with the naked eye; by 1870 it was below the sixth magnitude, and it has remained at about this level ever since, though with definite fluctuations. At one point it fell to magnitude 7.9, but has now recovered to just below 6.

I had my first view of it in 1968, when I was in South Africa. In Johannesburg there is a private observatory built and operated by Christos Papadopoulos, who is Greek (as you will gather from his name), but who has lived in South Africa for many years. The astronomical work which he carries out is equal to any professional standard, and he has produced a star atlas which is to be found in almost every major observatory in the world. Using his 15-inch reflector, he showed me Eta Carinæ. My immediate reaction was that Eta did not look like a normal star at all; it was more like what I recorded in my notebook as 'an orange-red blob'. I have seen it many times since, and my impression is always the same. It is also associated with a magnificent diffuse nebula, nicknamed the Keyhole Nebula from the shape of one of the dark areas in it.

With binoculars, you can find Eta Carinæ and its nebulosity quite easily. The only problem is that you will have to go to a location well

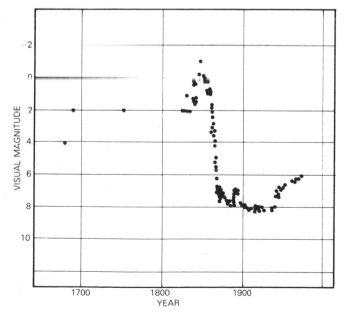

Fig. 53 Light-curve of Eta Carinæ

south of latitude 30 degrees north, as otherwise Eta will never rise sufficiently high above the horizon. The Canary Islands are too far north; Hawaii will just about do, but I recommend somewhere such as South Africa, Australia or New Zealand.

From the records, we can infer that Eta has always shown colour. In 1850, after it had started its decline, it was still described as 'reddish yellow, somewhat darker in hue than Mars'. When its spectrum was first photographed, in 1892, it seemed as though Eta must be a normal supergiant of type F, but three years later there had been a dramatic change. Instead of the usual dark lines of a stellar spectrum, Eta Carinæ showed bright lines, due to hydrogen, iron and other elements, but with a notable lack of oxygen, which would have been expected to be present.

Concerted efforts were made to find out how far away Eta Carinæ must be. The best method of attack was clearly to measure the distance of the associated nebula, whose official designation is NGC 3372. (NGC stands for the New General Catalogue of clusters and nebulæ drawn up by the Danish astronomer, J. L. E. Dreyer, though by now it is hardly new; it was completed practically a hundred years ago.) The nebula is lit up by hot blue stars, and the spectra of these give a distance of around 6000 light-years, so that the distance of Eta Carinæ is presumably much the same.

No bright naked-eye star is anything like so remote as this. Even Deneb, Rigel and Canopus lie well within 2000 light-years, and in 1843, remember, Eta Carinæ outshone them all. This means that if our distance estimates are right, Eta Carinæ in 1843 was some 6 million times as luminous as the Sun, making it the most powerful known in our Galaxy – or, for that matter, anywhere else.

There are two other important factors to be taken into account. First, most of Eta's radiation is sent out in the infra-red region of the electromagnetic spectrum, which does not affect our eyes; if we could 'see' infra-red, then Eta would be outstanding. Secondly, the star is enwrapped in nebulosity, and nebular dust is remarkably efficient at absorbing starlight, so that at the moment we may be observing Eta through a kind of cosmic fog. All in all, there is no reason to suppose that it is any less powerful now than it was in 1843, even though it does not look so brilliant to us. Moreover, it looks redder than it ought to do, again because of the effects of intervening dust. The surface temperature is of the order of 25,000°C, and the diameter could be around ten times that of the Sun – that is to say, 8–9 million miles (about 13–15 million km), which admittedly is not much when compared with a red supergiant such as Betelgeux.

Nothing seemed to fit in with the 1892 spectrum, which, as we have seen, led to the idea that Eta Carinæ was an ordinary F-type supergiant. But as has been pointed out by David Allen, of the Anglo-Australian Observatory, the 1892 spectrum may not have been that of the star itself, but of a shell of material which had been thrown off at the time of the great outburst of the early 1840s. The central star would be invisible, and the surrounding gas would have cooled to a modest 8000°C, about the same temperature as that of an F-type star. When the layer thinned out, we would see straight through to the real star, with its bright-line spectrum.

The extreme brilliance of Eta Carinæ at its peak was not due to closeness; as we have noted, it is a very long way away, and so far as is known there is nothing else like it within reasonable range. It is not an ordinary nova; it is far too luminous. Neither is it an ordinary variable star, because it does not look like one and does not behave like one. Therefore there are only a few possibilities.

First, we might be dealing with a true stellar infant, still preparing to settle down to a period of stable existence. But the high surface temperature seems to rule this out straight away, and in any case a very rarefied star radiating as powerfully as Eta Carinæ would not be able to remain intact. It would break up.

Secondly, could Eta Carinæ be a strange type of binary? We know that normal novæ are binary systems, and it is true that some of them are relatively slow to fade after their initial outburst. The slowest nova of modern times was discovered in 1967, in the constellation of Delphinus (the Dolphin) by the well-known English amateur, George Alcock, who has an impressive tally of novæ and new comets to his credit. At the time of the discovery, the nova – HR Delphini – was just about visible with the naked eye, and easy with binoculars. I think I was the first to confirm its existence, but I can claim absolutely no credit, because George Alcock telephoned me at a late hour and told me just where it was. The star remained of the fourth magnitude or brighter for some time, and even at the present moment (mid- 1987) it is still above magnitude 12. For once we have a record of it as it used to be before the 1967 blaze-up, and it seems to be declining to its original magnitude. Could something of the same sort apply to Eta Carinæ?

Again there seem to be fatal objections. We are again faced with the luminosity problem; we would have to assume that the white dwarf component of the binary suffered not a nova outburst, after which it would return to its old state, but a supernova explosion.

True, a Type I supernova, which involves the total destruction of a white dwarf, can produce at least three times the luminosity of Eta Carinæ at its peak, but the outburst could not persist for more than a year or two, and Eta has been observable for much longer than that, quite apart from the fact that over all wavelengths it is probably just as powerful now as it was in the 1840s. I suppose we cannot entirely rule out the chance that Eta Carinæ is a binary of exceptional type, with mass flowing from one member of the pair to the other, but it does seem extremely unlikely.

This leads to one more possibility: that Eta Carinæ is approaching a crisis in its evolution, and that at some time in the near future, on the cosmic scale, it will 'go supernova'.

This is the theory which has been developed by three American astronomers, N. R. Walborn, T. R. Gull and K. Davidson. What they have done is to make spectroscopic studies of the wisps of nebular material which were ejected by Eta Carinæ in the period from 1835 to 1843, and which have been moving outward ever since, so that they now lie 10 seconds of arc away from the star itself. Using the powerful reflector of the Cerro Tololo Observatory, in Chile, the three astronomers looked for indications of oxygen in Eta Carinæ's spectrum, but with no success. On the other hand, there was a great deal of nitrogen. Next they called in an Earth satellite, the IUE (International Ultra-violet Explorer), and hunted for traces of carbon. This too was absent, but again nitrogen was very much in evidence.

This was significant. In interstellar gas, oxygen is seven times as plentiful as nitrogen, and carbon four times as plentiful, so that evidently the wisps from Eta Carinæ were exceptional. But a solution was to hand. Inside very massive stars, nuclear reactions convert the original carbon and oxygen into nitrogen, so that in the long run there is a shortage of both carbon and oxygen. Eta Carinæ, then, is highly evolved, and it is also unusually massive. Generally speaking, stars are between one-tenth and thirty times as massive as the Sun; there are exceptions, of course, but there is a definite limit. At the moment the most massive star-system known is believed to be a binary in the constellation of Monoceros, the Unicorn, known officially as HD 47129 and more commonly as Plaskett's Star, because attention to it was first drawn in 1922 by the Canadian astronomer, John Plaskett; it is made up of two bright white supergiants, each with a mass fifty-five times that of the Sun, while the primary has a diameter of at least 20 million miles (32 million km). The two components are close together, with an orbital period of 14 days. Yet they are not in the same luminosity class as Eta Carinæ, and even when combined they could hardly match more than about 100,000 Suns.

It has been calculated that a star which is more than a hundred times as massive as the Sun will start to blow away its outer layers if it has 4 million times the Sun's luminosity – in which case Eta Carinæ may be over the theoretical limit, and will be unstable. Its mass is uncertain, but could well be at least the equal of the two components of Plaskett's Star put together.

122

We must look carefully at the way in which the star has behaved since detailed studies of it began. David Allen has suggested that as it condensed out of the nebular material it produced energy so prodigally that it has produced periodical outbursts ever since, ejecting its outer layers to form shells; one of these shells was responsible for the outburst of the 1840s, while the previous shell has almost disappeared, though faint traces of it were discovered years ago by David Thackeray with the aid of the 74-inch reflector at the Radcliffe Observatory in South Africa. At the moment the star is certainly losing mass, and it could well collapse to produce a Type II supernova. If so, it would be remarkably brilliant, and would illuminate the southern skies for months.

Of course, we have no real idea of when this will happen, if indeed it will happen at all; it could be in a few years, a few thousand years, or a few million years – we cannot tell. Meanwhile, astronomers tend to keep a close eye upon Eta Carinæ. With a supernova remnant such as the Crab, we may be quite sure that there will be no further outburst, because the old star has been transformed into a gas-cloud and what has become unromantically known as a 'stellar corpse'. But Eta Carinæ is not a supernova remnant – yet. It could blaze up once more at any time, and if it ejects a new shell, or if it emerges from behind its shield of obscuring dust, it may rival Sirius just as it did a century and a half ago. It is indeed a fascinating and unpredictable star.

XIII SS433: THE COSMIC LAWN-SPRINKLER

Most of the stars I have talked about so far are reasonably bright. Betelgeux is one of the most brilliant of all; in its heyday Eta Carinæ was even brighter; Sirius is in a class of its own, and even the modest 61 Cygni is easy to see with the naked eye. But I come now to a star which is very much dimmer. It lies in Aquila, the Eagle, but you need a telescope to see it. I have shown its position on the chart (Fig. 54) but, even if you located it, there would be nothing to single it out. It has no individual name; we call it SS433, because it was the 433rd entry in a catalogue of stars with unusual spectra, drawn up in 1877 by the American astronomers Bruce Stephenson and Nicholas Sanduleak. Yet it is now one of the most studied stars in the sky, and several international conferences have been devoted entirely to it.

Actually, the story of SS433 had its origins earlier, with some results from an obscure object in the far-southern constellation of Circinus, the Compasses. The object was listed as Circinus X–1, because it was a source of X-radiation; it appeared to be a supernova remnant, and it emitted radio waves as well, so that it was of special interest.

X-rays are of very short wavelength, and cannot be studied from the Earth's surface because of the shielding effects of our atmosphere. The X-ray region extends from 0.01 to 100 Ångströms, one Ångström being equal to one hundred-millionth part of a centimetre (10^{-10} metres).* If we are going to study X-rays from space, we must use either rockets or artificial satellites, and this has been possible only during the last twenty-five years.

---

*The unit is named in honour of the last-century Swedish physicist, Anders Ångström. It was inconsiderate of him to have a name beginning with the distinctive Swedish letter Å!

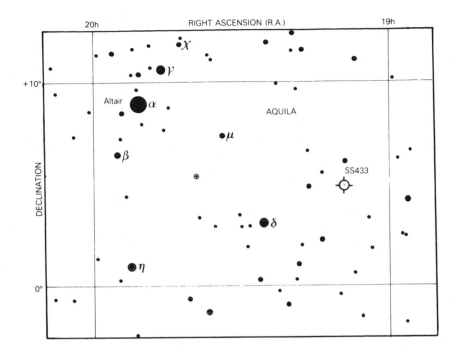

Fig. 54 Position of SS433

A very important artificial satellite, named Uhuru, was launched in 1970. It was equipped with an X-ray telescope, and it proved to be very successful. Obviously an X-ray telescope is very different from an ordinary telescope, but there is no need to go into details about it here; it is enough to say that Uhuru discovered many new X-ray sources, and achieved more than its makers could possibly have hoped.

One object to which Uhuru paid attention was Circinus X–1. Undoubtedly it was a discrete X-ray source, but it was also a radio emitter, and to make things even more interesting it was associated with what seemed to be a supernova remnant 100,000 years old (give or take a few tens of thousands of years either way!). But there was always the chance that Nature was playing a trick on us, and that the X-ray and radio sources were simply lined up with the supernova remnant as seen from Earth. It would clearly be a great help to find out whether there were any other objects similar to Circinus X–1. One promising candidate emerged: a supernova remnant in Aquila, known as W50. Near its centre was a small, fairly strong radio source which was definitely not a pulsar, and before long equipment carried on a British satellite, Ariel 5, showed

125

that there was also an X-ray source there. It was all most intriguing, and all that remained to do was to find a visible star in the same position – at least, so it was generally believed.

In June 1978 two English astronomers, Paul Murdin and David Clark, were using the Anglo-Australian Telescope at Siding Spring, New South Wales. This is a 158-inch reflector – not the largest telescope in the world but very possibly the best. Armed with information about the position of the radio source in W50, Murdin and Clark began their hunt. There were several faint stars in more or less the right area, but there was also one which was brighter. It had been expected that the W50 star (if it existed at all) would be dim, and so it was not for some time that the brighter object was examined – and, moreover, care had to be taken not to damage the highly sensitive equipment by swamping it with too much light. But when Murdin and Clark examined the spectrum of the brighter star, they realized at once that it was the object for which they had been searching.

First, there was the unusual spectrum. It included bright hydrogen lines, just as with Circinus X–1, which is why it had been included in the catalogue by Stephenson and Sanduleak (something which the British workers only found out later). But the strangest factor of all was that the spectrum was variable over short periods, with a longer cycle of 164 days. One group of lines was red-shifted and another blue-shifted, while yet another group showed no shift at all; some of the lines periodically vanished, only to reappear later. It was all completely chaotic, and unlike anything else known, so that Murdin and Clark did not know what to make of it. Assuming that the line shifts were Doppler effects, SS433 seemed to have the remarkable ability to approach, recede and stand still all at the same time!

The published results brought SS433 very much to the fore. Investigators in various countries tackled its unique problems, and it must be admitted that there was a certain lack of co-ordination between the different groups, because quick publication to establish priority became the order of the day. But gradually a more coherent picture started to emerge.

There was little doubt that the radio source, the X-ray source and W50 really were associated, and were about 10,000 light-years away from us. Secondly, W50 itself gave every impression of being a true supernova remnant. Thirdly, SS433 was not a single star; there were two components, moving round each other in a period of just over 13 days. The visible star was reddened, and very luminous, but the second component could not be detected at all, and thoughts of neutron stars or even black holes began to come into the theorists' minds.

126

The supernova remnant had to be explained, and was presumably due to the secondary component, because the visible star was still shining away quite happily. We have to concede that even today there is no final proof, but when asked whether he believed W50 to be a supernova remnant the Canadian astronomer, Sidney van de Bergh, one of the leading authorities on such matters, made a very sage comment: 'If it waddles like a duck, and quacks like a duck, then maybe it *is* a duck.' (Because W50 had originally been identified years before by an Indian astronomer, Thangasamy Velusamy, some people have even referred to it as 'Bombay duck'!)

A Type I supernova, which involves the destruction of a white dwarf, leaves no débris apart from faint wisps of nebulosity which emit radio waves, so that so far as SS433 is concerned we can rule it out. A Type II supernova produces a neutron star, and by the 'duck' analogy it seems that the secondary of SS433 might well be a neutron star. On the other hand, could it be a black hole?

Everything seemed to fall into place, but there are some dissentients, notably two Ohio astronomers, George Collins and Gerald Newsom, who reject the idea of a neutron star or black hole,

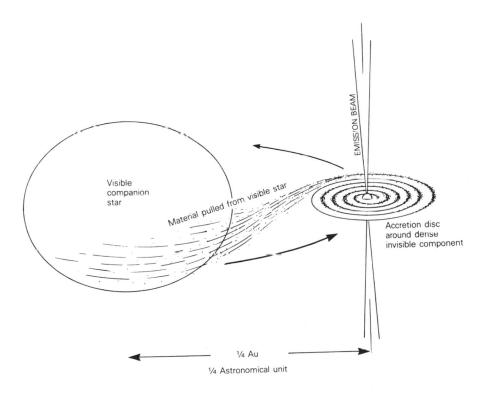

Fig. 55 Lawn-sprinkler theory of SS433

and believe that SS433 is made up of a very powerful blue supergiant which is being perturbed by a hot, massive companion and made to 'wriggle', propelling jets into space because of its powerful magnetic field. At the moment this is a minority view, though certainly it cannot be ruled out.

On the more generally-accepted picture, SS433 would indeed be remarkable if it could be seen from close range, with its luminous primary, its condensed companion and its wobbling jets. Even if the companion were a black hole, the view would still be extraordinary, because there would be the swirling, intensely heated material which has been pulled off the visible star and is emitting X-radiation just before being pulled over the event horizon of the black hole and being lost to the outside universe for ever.

The real lesson of SS433 is that it has shown us that despite all our recent advances in theory and observation, there are still many classes of objects in the Galaxy about which we not only know little, but do not even suspect. Twenty years ago, who would have thought that a dim, undistinguished-looking star in the middle of an obscure patch in the Eagle would turn out to be the fascinating system we have since found it to be? Whether we will manage to detect any more 'cosmic lawn-sprinklers' remains to be seen, but no doubt they exist.

XIV ALCYONE: BRIGHTEST OF THE SISTERS

Look into the eastern sky soon after dark in what to Europeans is late autumn and to Australians late spring, and you will see what looks at first sight to be a filmy mass. Examine it more closely, and you will see that it contains stars, one of which is decidedly brighter than the rest. See how many stars in the patch you can count. Under good conditions you should manage at least seven, though people with excellent eyesight may raise the total to ten or a dozen, while the record is said to be nineteen. The stars in the patch are not lined up by chance. They form a cluster – the finest loose or open cluster in the sky, known to astroners as M45 or the Pleiades, and to most laymen as the Seven Sisters (Fig. 56).

Because the Pleiades are so conspicuous, they have been known from very early times. In Homer's *Odyssey*, Odysseus 'sat at the helm and never slept, keeping his eyes upon the Pleiades'. In China

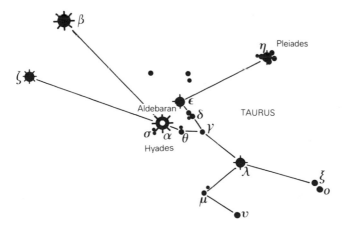

Fig. 56 Position of the Pleiades

it is said that they were recorded as early as BC 2357, and were worshipped by girls and young women as the Seven Sisters of Industry (which sounds more like modern Soviet Russia!) while the Hindus pictured them as representing the flame symbolic of their fire-god Agni. Hesiod called them the 'Seven Virgins'; Manilius, in the first century BC, referred to them as the 'Narrow Cloudy Trail of Female Stars'. They even feature in the Bible. In the Book of Job we find the famous lines: 'Canst thou bind the sweet influence of the Pleiades, or loose the bands of Orion?' And it is Orion who features in a Greek myth about the cluster.

According to the story, the Pleiades were seven beautiful girls who were wandering in the woods when they were spied by Orion, who was a mighty and fearless hunter, but who had the usual human emotions. As soon as he saw the girls he gave chase, with intentions which were clearly anything but honourable. The Pleiades fled; Orion thundered after them; from Olympus, the king of the gods, Zeus (better known to us as Jupiter) intervened, saving the maidens' virginity by transforming them into stars and placing them in the sky.

Orion's feelings about the whole episode are not on record, but certainly the constellation which represents him can be used as a guide. Follow up the line of the Belt stars until you come to the bright orange-red Aldebaran, in Taurus.* Continue the line, curving it somewhat, and you will reach the Pleiades. Note, however, that the Pleiades are further west, so that they rise before Orion. From my home in Selsey, in southern England, I always feel that the first sight of the Pleiades in the evening sky means that winter, with its fogs and frosts, it not too far away.

Both Aldebaran and the Pleiades lie in the Zodiacal constellation of Taurus, the Bull. The brightest of the Seven Sisters, Alcyone, is officially known as Eta Tauri. It is just above the third magnitude, so that it is somewhat fainter than the main stars of the Plough.

Alcyone may be the senior member of the Pleiades, but it is not so important as some people believed during the last century. There was, in particular, the German astronomer, Johann Heinrich von Mädler, who was a brilliant observer; in collaboration with his friend and pupil, Wilhelm Beer (brother of Meyerbeer, the composer), he produced a map of the Moon which was by far the best of its time. It was published in 1838, and was a true masterpiece even though it was compiled with only a small telescope – the 3¾-inch refractor in Beer's observatory outside Berlin – and of course without the aid of photography. Mädler then left Berlin to become director of the new

---

*From northern latitudes, that is to say. From the southern hemisphere the Belt stars point *downward* to the Pleiades and *upward* to Sirius.

observatory at Dorpat in Estonia, which was then, as now, under Russian control. Mädler devoted his time at Dorpat to stellar astronomy rather than lunar work, and he was notably less successful. He put forward the strange theory that Alcyone marked the true centre of the Galaxy, and that all the other stars, including the Sun, revolved round it.

I have never understood how Mädler came to this conclusion, because there seemed to be absolutely no reason for it. It was not widely accepted, but it was still repeated in many popular books until well into the second half of the nineteenth century. Today we relegate Alcyone to the status of a hot bluish star, around 1000 times as powerful as the Sun, lying at a distance of 410 light-years. It is large, with an estimated diameter of between 8 million and 9 million miles (13–15 million km), and it is comparatively young. The name 'Alcyone' is of course that of a Greek maiden, though the early Arabs called it, less romantically, 'Al Jauz' or 'the Walnut'.

We cannot discuss Alcyone on its own; its story is bound up with those of the other Pleiades stars (Fig. 57). Of these Electra, Atlas, Merope, Maia, Taygete and Pleione are visible with the naked eye, though Pleione lies close to the brighter Atlas and the two are not always easy to separate. Next come Celæno and Asterope, which are on the fringe of naked-eye vision. Binoculars bring many more into view, and the total membership of the cluster amounts to several hundreds. (Atlas is the only Pleiad with a male name. He was, apparently, the husband of Pleione and father of the sisters.)

Before going any further, we must make sure that we really are dealing with a genuine cluster; remember the Milky Way, whose

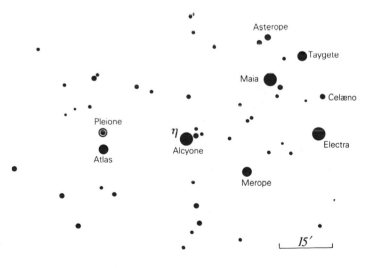

Fig. 57 Diagram of the main stars in the Pleiades

stars are apparently so packed and yet have no real connection with each other. All we have to do is to make some very rough calculations. The cluster is about 120 minutes of arc across, which is four times the apparent diameter of the full moon. Work out the odds against all the Pleiades being crowded into this area by sheer chance, and you can see that any such idea is completely out of court; we also know that the members of the cluster share a common motion in space, and are at approximately the same distance from us (slightly over 400 light-years). There are, of course, a few line-of-sight effects, and two stars of magnitude 7.5 are definite intruders, because their distances and proper motions are quite different. But over 98 per cent of the stars in the area are genuine cluster members.

The other open cluster in Taurus surrounds Aldebaran, the orange-red 'Eye' of the Bull, and is known as the Hyades (Fig. 58). The arrangement is not the same as with the Pleiades. The Hyades are brighter on average, and extend away from Aldebaran in a sort of V-formation, though Aldebaran itself is not a cluster member, and merely lies midway between the Hyades and ourselves. Incidentally, the Hyades are closer than the Pleiades, and are in fact the nearest of all the open clusters.

There is a minor mystery about the number of naked-eye stars to be seen in the Pleiades, and there has been talk of a 'missing

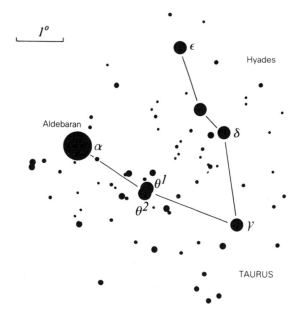

Fig. 58 Diagram of the Hyades

maiden', so that one of the Sisters has vanished or at least faded considerably since ancient times. The Roman writer Ovid even stated: 'Quæ *septem* didi, *sex* tamen esse solent'. The magnitudes of the leading members are:

Alcyone:	2.9	Taygete:	4.3
Atlas:	3.6	Pleione:	5.1
Electra:	3.7	Celæno:	5.5
Maia:	3.9	Asterope:	5.8

If we merge Pleione with Atlas, or regard Asterope as being below the limit, we are indeed left with seven, but to normal eyes Pleione and Atlas can be separated. Either way, there is no reason to reduce the accepted number to a mere six. To test this out, I conducted an experiment some twenty years ago during one of my *Sky at Night* programmes on television. I talked about the Pleiades, and then asked viewers to write in to tell me how many stars they could see without optical aid on a dark, clear night. I also asked for a rough map, to make sure that the helpers were looking at the right area. There were over 5000 replies, and the results were fairly conclusive. The maximum number from a reliable viewer was twelve; the lowest five; the average worked out at precisely 7. (I had to disregard some honest but erroneous replies; one lady claimed to see over 150 stars, but it was clear from her map that she had counted not only the Pleiades but also the rest of Taurus, and had even thrown in part of Orion for good measure!)

On the other hand one of the main stars, Pleione, is variable. It has a range of about half a magnitude, which is not very much, but which may take it alternately above and below naked-eye visibility. It is what is termed a 'shell star', and is unstable; at times its spectrum indicates that it throws off gaseous shells which make it brighten temporarily until the shell dissipates. This happened, for example, in 1937–8 and again in 1972. If there is any truth in the legend of the 'lost Pleiad' it seems likely that Pleione is responsible, but personally I am dubious. We cannot place complete faith in the old records.

There are no red giants in the Pleiades, and this alone is enough to show that by cosmic standards the cluster is young; there has not been enough time for the more massive stars to use up their hydrogen fuel and leave the Main Sequence. We are dealing with a typical Population I area, and, as might be expected, there is a great deal of star-forming material left. Photographs show beautiful extended nebulosity, brightest in the region round Merope. The Pleiades nebula is not easy to see at the eye-end of a telescope, though, rather surprisingly, it was discovered as long ago as 1859 by the German astronomer, W. Tempel, with a refractor of only 4 inches aperture. I can see it easily enough with my 15-inch reflector, and smaller instruments will show it, but one has to look carefully for it. It is a 'reflection nebula', shining by the light of the stars mixed in with it, and therefore different from an emission nebula such as the Sword of Orion, about which I will have more to say in Chapter 15. Very possibly the Pleiades nebula is variable in brilliancy, though no definite changes in its form or structure have been proved since the first good picture of it was taken just over a century ago.

Despite the lack of red giants, there are some very dim red stars in the Pleiades, believed to be true infants still condensing from the nebular material; their nuclear 'fires' have yet to be triggered off.

The whole cluster is drifting through space, covering an apparent distance of 5.5 seconds of arc per century, so that it will take 30,000 years to cover a distance equal to the apparent diameter of the full moon. This corresponds to a true relative velocity of 25 miles per second (40 km per sec.).

If we assume that all the main stars in the Pleiades were born in the same way, at around the same time, and from the same cloud of material, we can make at least a reasonable estimate of their age, and here again we can take Alcyone as a guide – because Alcyone is the brightest member of the cluster, so that it must also be the most luminous and, presumably, the most massive. We can calculate how long it would take Alcyone to use up its available hydrogen and leave the Main Sequence. The answer is well below 100 million years, and so the Pleiades must be younger than this. When we take the less massive members of the cluster into account we also obtain an absolute upper limit of 100 million years, and with Alcyone the true value is likely to be nearer 20 million years. Moreover, there is the unstable condition of Pleione; can it be that here we have a star which still shows the signs of extreme youth?

Obviously there are fundamental differences between open clusters such as the Pleiades and globular clusters such as Omega Centauri. The globulars are old, and of Population II, with their leaders well into the red giant stage, and they are much more populous; a typical globular may contain at least a million members as against only a few hundreds for even a rich open cluster such as the Pleiades.

Of course, not all the open clusters are of the same age; the Hyades are more evolved than the Pleiades, for example, but the overall pattern is always much the same. Look at other open clusters – Præsepe or the Beehive in Cancer, the lovely Jewel Box in the Southern Cross, the Wild Duck cluster in the Shield, and so on. They are less spectacular than the Pleiades because they are farther away, but they are of the same basic type. Even one exceptionally evolved cluster, M67 in Cancer (not to be confused with Præsepe) is much younger than the average globular.

We cannot regard open clusters as being permanent features. The stars in them may have a common origin, but they are only loosely 'bound', and they are certain to be affected by the gravitational pulls of non-cluster stars in their vicinity. Moreover a cluster member may, by chance, acquire enough relative velocity for it to break free and be lost to the group. Over a sufficient period of time, an open cluster will become so scattered that it will cease to be identifiable. The stars will have more or less the same velocity and direction of motion through space, but they will have changed into what is called a 'stellar association' or moving cluster. This is true, for instance, of five of the stars in the Plough, and this particular association also includes Sirius.

If we lived upon a planet orbiting one of the stars in the Pleiades, we would have a brilliant night sky. In the cluster there are several hundreds of stars contained in a sphere about 50 light-years in diameter, and some of these are much more luminous than the Sun. Of the fifty brightest stars in our own sky, there are only eleven which are closer than 50 light-years – Alpha Centauri, Sirius, Arcturus, Vega, Capella, Procyon, Altair, Pollux, Fomalhaut and Castor – and of these only Arcturus and Capella qualify as giants over a hundred times the Sun's luminosity. Not one of them is comparable with Alcyone, or even with the other naked-eye Pleiads. There would also be the widespread hazy nebulosity, so that the sky would never be black.

Whether there are any civilizations in the Pleiades cluster seems, frankly, rather improbable, because the cluster itself is too young, but we cannot be sure, and there may be dwarf stars of solar type about which our ignorance is complete.

Photographs show the Pleiades in all their glory, but if you use a telescope you may feel disappointed, because the cluster covers too wide an area to be held in the same field even with a moderate magnification. With, say, Præsepe or the Jewel Box you can see the whole cluster at once; with the Pleiades you cannot, unless you have a rich-field instrument and use a very low power. In my opinion, the best view of the Pleiades is to be obtained with binoculars. With a 7 × 50 pair, or the equivalent, you will see all the Pleiades at the same time – not only the main pattern, but also many fainter stars. You will certainly have no trouble in recognizing Alcyone, which is so clearly the most brilliant of the Sisters. It is a lovely star, even though Johann Mädler was wrong in believing it to be the supreme sun of our Galaxy.

XV BECKLIN'S STAR: THE STAR WE CAN NEVER SEE

Of all the gaseous nebulæ the best-known, and the brightest as seen from Earth, is M42 in the Sword of Orion (Fig. 59). On a clear night you cannot possibly mistake it. The Sword extends away from the three stars of the Hunter's Belt – downward if you are observing from Europe, upward from Australia or South Africa – and it is clearly 'milky', with one brighter star at the far side. The brighter star is Iota Orionis, sometimes still called by its old proper name of Hatysa, which is at least 20,000 times as luminous as the Sun and is over 1500 light-years away. The Nebula lies between Iota and the Belt.

Though the Nebula is so prominent, it was not recorded *as* a Nebula until an otherwise obscure astronomer named Nicholas Peiresc did so in 1610, a few months after the invention of the telescope. Galileo himself had mentioned the region, and commented that he had 'added eighty other stars recently discovered' there, but there was no mention of actual nebulosity. Neither was it described as such by the famous observers of pre-telescopic times; even Tycho Brahe overlooked it (though it is true that he also overlooked the Andromeda Spiral, which is definitely visible with the naked eye even though it is much fainter than M42).

Recently there have been suggestions that the Nebula was once genuinely fainter than it is now, which would explain why it was not mentioned in pre-telescopic times, but I take this with a very large grain of cosmic salt. Certainly it is true that the Nebula shines because of the stars associated with it, and if these stars brightened up then the Nebula would brighten too; but to suppose that this happened at the very time when Earth astronomers had built their first telescopes to observe it is stretching coincidence rather too far!

The stars chiefly responsible for making the Nebula luminous are those of the multiple star, Theta Orionis, often referred to as the

136

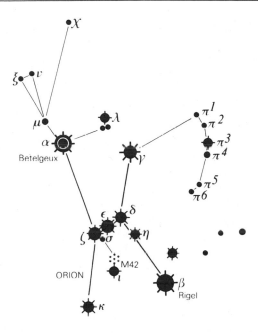

Fig. 59 Position of the Orion Nebula

Trapezium for reasons which are obvious as soon as you look at it. But of even greater interest is a star inside the Nebula which is hidden from us. We know it as the Becklin-Neugebauer Object, or BN for short; nobody has ever had a direct view of it, and nobody ever will, but we are fairly sure that it is a huge, immensely powerful star, pouring out its radiation at a furious rate only to be effectively blocked by the nebular material.

First, then, a few words about nebulæ in general. We have already noted that they are stellar birthplaces. In the sky they are common enough, and quite a number were included in Messier's catalogue. They are of two main types. The gas in each type is mainly hydrogen (after all, hydrogen is much the most plentiful element in the universe), and there is also 'dust', which is very efficient at blotting out starlight. With reflection nebulæ, such as that in the Pleiades, the nebular material simply reflects the light of the stars, and the hydrogen atoms remain complete – each atom having a nucleus around which orbits a single electron, so that the atom itself is electrically neutral. (Remember that a proton, which makes up a hydrogen nucleus, has unit positive charge, while an electron has unit negative charge; $+1 - 1 = 0$.) But if the stars in or very near a nebula are extremely hot, they break up or 'ionize' the hydrogen atoms because of their radiation, so that the broken atoms or 'ions' emit a certain amount of luminosity on their own account.

137

Nebulæ of this kind, such as that in the Sword of Orion, are termed emission nebulæ. Astronomers prefer to give a different classification: HI regions where the hydrogen atoms are complete, HII regions where they are ionized. The Sword of Orion is a typical HII region.

We also know many dark nebulæ, which were described by the great observer William Herschel as 'holes in the heavens'. For once Herschel was wrong. The dark nebulæ are not voids; they are just the same as shining nebulæ, apart from the fact that they are not associated with any suitable stars to light them up. The best example is the Coal Sack in the Southern Cross, unfortunately never visible from Europe. With the naked eye you can notice an almost starless patch, which is simply a dark mass cutting out the light of stars beyond. Yet we can only see the side of it which is turned toward us; for all we know, on the far side there may be a powerful star illuminating it – so that if we were observing from a different vantage point in the Galaxy, we might see the Coal Sack shining more brilliantly than the Sword of Orion.

The Orion Nebula contains large numbers of faint stars which vary in light quite rapidly and irregularly. They are known, rather misleadingly, as T Tauri stars, because the first-known member of the class was found not in Orion, but in the adjacent constellation of the Bull. Undoubtedly they are very young indeed, and are still condensing; they have not yet started their nuclear 'engines', so that they are flickering as they shrink toward the Main Sequence in the HR diagram. No doubt the Sun, too, went through a T Tauri stage, and there is excellent evidence that it did – the effects upon its family of planets, including the Earth, were quite dramatic.

Star formation in the Nebula is still in progress, and there are several cases in which stars have been observed to 'switch on', as it were, blowing away their dusty shells so that we can see them. There are also a few stars which are nebulous, and appear to be speeding away from the Sword in opposite directions at a velocity of around 80 miles per second (130 km per sec.); two of them are called Mu Columbæ and AE Aurigæ. They are now over 1000 light-years from the Nebula, and it is thought they were born there no more than 3 million years ago.

The visible Nebula is only part of a huge cloud which covers most of the constellation of Orion. It is not detectable in visible light, but there are other methods about which I will have more to say shortly.

The distance of the Orion Nebula is 1600 light-years, so that we now see it as it used to be when the Romans were preparing to abandon Britain. The visible part has a diameter of 16 light-years; that is to say, if we happened to lie in its centre Sirius would be just at the outer edge. Despite its vast size, it is not so substantial as might be thought, and the gas probably accounts for no more than 200–300 times the mass of the Sun. It is very rarefied, with around 600 atoms per cubic centimetre, which compares with the best vacuums which we can produce in our laboratories. Calculations made some time ago, but not significantly revised, indicate that hydrogen accounts for more than 90 per cent of all the atoms; and it is hydrogen out of which stars are born.

To show how tenuous the material is, David Allen once told me that he had calculated the 'weight' of the material which would be collected by an inch core sample right through the Nebula from one side to the other, a distance not much less than 100 million million miles. The total weight would be no more than that of an ordinary pencil.

The four main stars of the Trapezium are hot enough to ionize the hydrogen atoms in their neighbourhood, and two of them are short-period eclipsing binaries. There is no doubt that all the Trapezium stars are genuinely associated, and have a common origin; they also mark the core of a compact cluster of faint stars whose members may possibly be spreading outward from the condensed region in which they were born. Some recent calculations indicate that the start of the expansion of this particular cluster began less than 300,000 years ago, making it so young that even the Pleiades seem like senior citizens by comparison. We also see small, bright condensations which are called Herbig-Haro objects after the astronomers who first drew attention to them, and which are, presumably, embryo stars.

If the Trapezium stars lay deep inside the Nebula we could not see them, powerful though they are. It is not the gas which causes the obscuration; it is the 'dust', and even in so rarefied a mass the solid particles conceal objects beyond them. The Trapezium stars lie very near the Earth-turned edge of the Nebula. If they are about 100,000 years old, which seems a reasonable estimate, they have had enough time to 'burn a hole' in the dark cloud, so that they have blown away sufficient material to allow their light to escape. Had they been farther inside this would not have been the case, and this brings me on to the mysterious hidden object BN.

Visible light cannot percolate through from the heart of the Nebula, but infra-red can. It is less easily stopped by the obscuring material, but unfortunately it is absorbed by our own atmosphere, particularly by water vapour, and so infra-red astronomers are handicapped from the outset. To make the best of matters, infra-red telescopes have been set up at high altitudes. The largest British telescope of this sort is UKIRT – the United Kingdom Infra-Red

Telescope – which has a 150-inch mirror, and has been put on top of the 14,000-foot extinct volcano Mauna Kea, in Hawaii.*

Because infra-red radiations are longer in wavelength than visible light, it was thought – correctly – that the mirror for an infra-red telescope need not be of such high accuracy. This meant that it could be thinner, and, consequently, cheaper; the mounting, too, could be comparatively lightweight. UKIRT cost much less than a conventional telescope of the same size would have done; but as soon as it had been tested, astronomers realized that it was good enough to be used for normal work as well as infra-red, which was sheer bonus. Mauna Kea remains the highest major observatory in the world, and many telescopes are now to be seen on or near the summit.

Even before the Mauna Kea observatory had been established, two American astronomers, Eric Becklin and Gerry Neugebauer, had discovered an important infra-red source inside the Orion Nebula (hence the abbreviation BN). Though we can pinpoint its position with reasonable accuracy, we cannot 'see' a trace of it; we rely entirely upon its infra-red radiation, and do our best to decide just what it is. It could be a group of fresh stars, still in the process of condensing; but in this case, would the emission be as strong as it actually is? BN cannot be a nova or supernova; it does not vary noticeably, and it has lasted for years since its discovery, which rules out any nova-like catastrophe. The other alternative is that it is simply an extremely powerful star, and that all its radiation at optical wavelengths is trapped inside the Nebula.

From the strength of the infra-red, we can deduce that BN is at least a million times more energetic than the Sun, and may even be comparable with that cosmic freak, Eta Carinæ, though so far as we can tell it is much less erratic in his behaviour. Unlike Eta Carinæ, however, BN cannot be very old – a million years or so seems likely – and neither can it persist for long in its present form. Given enough mass, it could well produce a Type II supernova, but for all we know it may be able to shed its excess mass before such a disaster becomes imminent.

It is not the only known infra-red source inside the Orion Nebula. There are others, of which the most notable is KL or the Kleinmann-Low Object (again named after its co-discoverers). But KL seems to be different from BN, and is definitely not identical with it; apparently there are several centres from which material is

---

*At least, one hopes that Mauna Kea is extinct. Its neighbour, Mauna Loa, is active, and on one occasion during the last century lava from it menaced Hilo, the only real town on the island. According to legend, the lava was stopped at the critical moment by the intervention of a local witch-doctor, who had been hastily summoned to come to the rescue!

expanding quickly, in which case KL is likely to be a cluster of proto-stars rather than one super-supergiant.

At the moment this is about as much as we can find out, but it is fascinating to speculate as to what we might see if we could venture into the Orion Nebula. We would pass the hot stars of the Trapezium, and then plunge into the vast though rarefied cloud; eventually our view of the outer Galaxy would be cut off, and only occasional dim stars and condensations of gas and dust would be seen. Finally we might begin to see the glow of BN, not by its infra-red but by its ordinary light; and if we could go close enough, we would see it in all its glory – brighter than a million Suns, strongly reddened by the dust around it, and using up its reserves of energy at a rate which must, in the foreseeable future, destroy it.

Whether this idea of BN is correct, or whether we are dealing with something quite different, remains to be seen. But it does appear that we are on the right track, and we can at least visualize the scene near BL – one of Nature's searchlights, doing its best to penetrate its surrounding fog and make its presence known, but failing even in spite of its tremendous power. Unless there are intelligent beings living on planets orbiting stars inside the Sword of Orion – and I can think of few things more unlikely – BL is a star which will live and die totally unseen.

XVI S ANDROMEDÆ: THE NEW STAR IN THE SPIRAL

It might not be thought that there could be much of a connection between the Andromeda Spiral (Fig. 60), a huge system of stars over 2 million light-years away, and a house-party given on 22 August 1885 by a Hungarian baroness named de Podmaniczky – but there is. The baroness had a marked interest in astronomy, and one of her guests on that particular occasion was a professional scientist, Dr de Kövesligethy. To interest the other guests, they took out a modest 3½-inch refractor belonging to the baroness, and started looking around the night sky. One of the objects on view was the Spiral. When the baroness focused the telescope, she commented that she could see a star shining in the middle of the filmy mass. Dr de Kövesligethy agreed, but he believed that the appearance was due to the presence of the Moon, which overpowered the fainter parts of the nebula. The telescope was duly packed up and put away, and the whole episode was forgotten – until there came exciting news from France, Russia and Germany.

At Rouen, Dr Ludovic Gully had made the same sort of observation on 17 August, almost a week before the baroness's house-party. He did not attribute the 'star' to moonlight, but to a fault in his telescope. On 20 August the star was seen by Dr Hartwig, who was on the staff of the Dorpat Observatory in Estonia (which, as you may recall, was the observatory from which Struve had made his pioneer measurements of the parallax of Vega almost fifty years earlier). Hartwig dutifully reported his discovery to the Observatory director, and met with an immediate rebuff. The director had not himself seen the star; the Moon was in the sky; there could be some mistake, and so no official announcement must be made. Hartwig, presumably annoyed, wrote to the centre of astronomical information at Kiel, but apparently the letter never arrived, and nothing more happened until 25 August, when the star

Fig. 60(a) Impression of S Andromedæ as it appeared in 1885

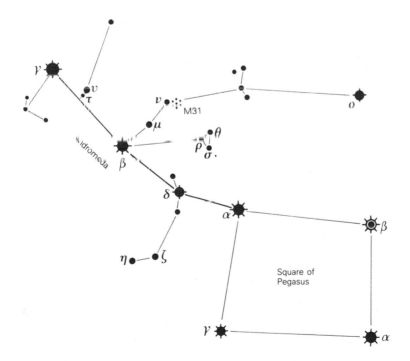

Fig. 60(b) Position of M31, the Andromeda Spiral

143

was seen by Max Wolf from Heidelberg. Wolf repeated the observation on the 27th; later, the Irish astronomer Isaac Ward claimed to have seen it on the 19th – but there could be a mistake here, because Ward gave the magnitude as below 9, which we now know to have been very much too faint. Finally, when the Moon was out of the way, Hartwig reobserved the star and showed it to the director. This time the news was released.

The failure to make a prompt announcement would be impossible today. It was regrettable in 1885, but it must be added that nobody realized the real significance of the strange star, and it was generally believed to be a variable or even an ordinary nova which happened to lie in that direction. For that matter, nobody knew the true nature of the Spiral either – and I have in my possession a book written in 1909 by a well-known astronomer and cosmologist, J. Ellard Gore, who gave the distance of the Spiral as 19 light-years! The older ideas of 'island universes' or separate galaxies, which had earlier been tentatively proposed by William Herschel, had been generally rejected, and were not to be revived in earnest until the Cepheid variables came to the rescue.

We are now sure that the star, which we know today as S Andromedæ, was no ordinary nova. It was much more luminous than that. Until recently it was believed to have been a supernova, but there have been doubts raised during the past few years, and it now seems more likely that S Andromedæ was more akin to our own Eta Carinæ. At any rate, it remains the only object of its kind to have been seen in the Spiral (though there have been plenty of normal novæ), and it has long since disappeared. Apparently the last observation of it was made on 1 February 1886 by Asaph Hall, discoverer of the satellites of Mars, with the 26-inch refractor at Washington. S Andromedæ was then of magnitude 16, and there was no hope of following it any further. The peak brightness was around magnitude 5.4, just within naked-eye range.

Nature can sometimes be very unkind. S Andromedæ flared up a century too soon. Had it appeared in 1985, not 1885, it would have been closely studied from every major observatory in the world; all facets of its behaviour would have been recorded and analysed, and we would have taken full advantage of the rise and fall of a colossal outburst only a little over 2 million light-years away. In 1885 nothing of the sort was possible, and I remember reading a comment by one experienced observer that, in his view, 'the star had nothing to do with the nebula'.

Had it been a normal nova, it would not have attained anything like the brilliance it actually did – and normal novæ are not too uncommon in our Galaxy. There was a particularly brilliant example in 1901, in the constellation Perseus, which was discovered by an amateur astronomer named Anderson and quickly became one of the brightest stars in the sky, though it soon faded. Max Wolf took some photographs of it a few months later, and found that the nova was then surrounded by nebulosity which appeared to be expanding at the speed of light. This, obviously, was most unlikely, and Wolf correctly decided that already-existing nebulosity was being steadily illuminated by the burst of light from the nova. After two years the nebula had become so large, and so faint, that it could be followed no further; by that time its apparent diameter was greater than that of the full moon.

By measuring the apparent expansion of the nebula, and knowing the velocity of light, it was possible to calculate that the distance of the nebula – and, hence, of the nova – was around 500 light-years. If it had been sixteen times further away than this, Nova Persei would have reached the same maximum brightness as S Andromedæ. Therefore, it was reasoned, S Andromedæ and the Spiral could be no more than about 8000 light-years away – assuming that the two outbursts peaked at the same luminosity. This estimate was a marked improvement on Gore's 19 light-years, but it still meant that the Spiral would be a minor feature in the Milky Way system rather than an independent galaxy. Opponents of the so-called 'island universe' theory (including, at that time, Harlow Shapley) made the most of this apparent proof.

But before long, uneasy doubts arose. Over the next years several new novæ were found, not only in the Andromeda Spiral but also in other 'starry nebulæ'. None of them could be compared with S Andromedæ; they were not only fainter, but enormously fainter. It began to look as though the baroness' star were a genuine freak.

When Edwin Hubble used the Cepheids to establish that the Spiral really does lie well beyond the edge of our Galaxy, the unusual nature of S Andromedæ became evident. Hubble himself wrote in 1929 that it belonged to 'that mysterious class of exceptional novæ which attain luminosities that are respectable fractions of the total luminosity of the system in which they appear'. Checks back through the old records put the stars of 1572 and 1604 into the same category (though not much was then known about the 1054 event), and the term 'supernova' was coined; it seems to have been first used by the Swiss astronomer Fritz Zwicky in the early 1930s.

Zwicky became fascinated by supernovæ. It would be too much to hope that one would conveniently appear in our Galaxy, and just as over-optimistic to believe that the Great Spiral would produce another S Andromedæ. So Zwicky decided to search in other galaxies, using equipment obtained specially for the purpose. His whole approach aroused a great deal of scepticism, and many of his colleagues said openly that they regarded the whole project as a

145

waste of time. Not to be daunted, Zwicky managed to obtain a new type of photographic telescope, a 'Schmidt', which uses a spherical mirror and a special type of correcting plate, and which can photograph a relatively wide area of the sky with a single exposure, whereas a conventional photograph with a large telescope has an extremely small field.* Between 1936 and 1975 Zwicky discovered a grand total of 122 supernovæ, and other observers added many more. It seems that on average, a galaxy will produce a supernova bright enough to be detected from Earth once in around 4–500 years. There must be many more which we do not see, because of intervening material – and this also applies to our own Galaxy.

Because supernovæ are so powerful, they can be seen over tremendous distances, and can be used as 'standard candles' in much the same way as the Cepheids, though they are less reliable. We can assume that all supernovæ attain much the same peak luminosity, which is not actually quite true, but was once believed to be so. We can then estimate the distance of the supernova, and, hence, of the galaxy in which it lies. It means waiting for a convenient supernova to come along, but there are plenty of galaxies from which to choose.

However, we now know that there are two distinct types of supernovæ. Those of Type II involve the collapse of a massive star, which ends up as an expanding gas-cloud plus a neutron star or pulsar – witness the Crab Nebula remnant. But with Type I supernovæ we are back to binary systems, and the whole story is most intriguing.

There are two components of the binary, one of which is more massive than the other and, therefore, evolves more rapidly (Fig. 61). It turns into a white dwarf, very rich in carbon, while the second component remains on the Main Sequence. The white dwarf pulls material away from its companion, and acquires a layer of hydrogen – but it has been found that no white dwarf can be more than about 1.4 times as massive as the Sun (the so-called Chandrasekhar Limit, named after the Indian astronomer who first pointed it out). As the white dwarf approaches this limit, it heats up to a 1,000,000,000°C or so, and catastrophe follows. This is a sudden nuclear explosion – and in a matter of a couple of seconds the carbon is changed into other materials (oxygen, neon and magnesium) in a chain of events which ends up as nickel. It is this which causes a Type I supernova. The white dwarf is utterly destroyed.

---

*Bernhard Voldemar Schmidt, inventor of this type of telescope, was a curious character; in his youth he blew off part of his right arm during experiments with gunpowder, and he was very much of what we would now call a 'loner'. He died, sadly, in an asylum for the insane; but his great contribution to astronomy, the Schmidt camera, is of tremendous importance in modern research.

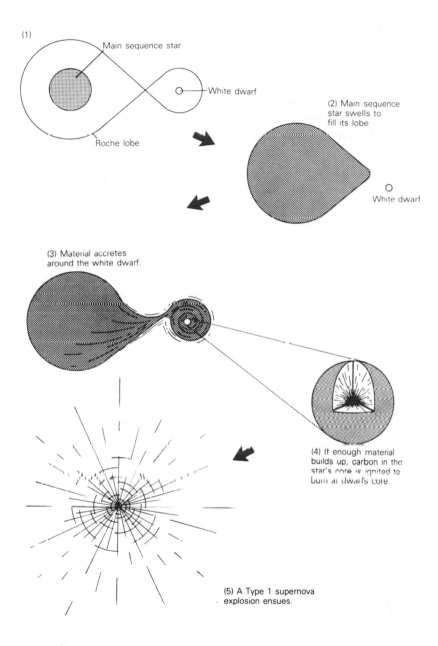

Fig. 61 Production of a Type I supernova

How can you tell one type from the other? Well, first there are the spectra; a Type II supernova contains plenty of hydrogen, a Type I has none. Secondly, the light-curves are different. A Type I fades rapidly at first, and then declines more or less regularly; a Type II is slower to drop in brightness, and the subsequent decline is less smooth. Moreover, the peak luminosity of a Type I is greater than that of a Type II. In some respects S Andromedæ was misleading, because it seems to have been underluminous for either type – which is why it is now more generally ranked with Eta Carinæ. But in estimating the distances of far-away galaxies, true supernovæ are of immense importance.

Clearly we want to detect supernovæ as early as possible, so that their light-curves can be drawn and their spectral changes analysed (Fig. 62). Zwicky showed the way; he has been followed by many other professional astronomers, and today amateurs are joining in. In England, R. W. Arbour has his 20-inch reflector fully computerized; once he has started his night's programme, the telescope will swing automatically from one galaxy on his list to another, taking a photograph of each which will be good enough to show up

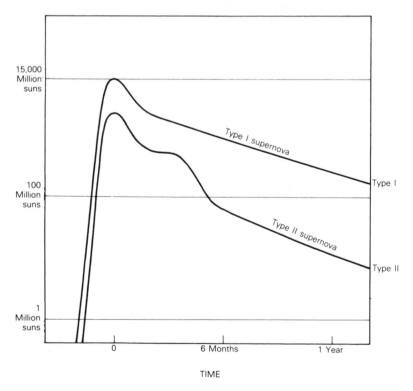

Fig. 62 Light-curves of Types I and II supernovæ

148

anything unusual. The Australian amateur the Rev. Robert Evans has a different method; he uses sheer observation at the eye-end of his 12-inch reflector, and he has memorized the positions and appearances of all the galaxies which he has selected. The advantage here is that he can detect any outburst immediately, whereas the photographers have to wait to develop their pictures. To date (1987) Evans has discovered twelve external supernovæ, while two other amateurs, Jack Bennett of South Africa and G. Johnson of the United States, have found one each.

On 3 May 1986 Evans had one of his successes. He detected a supernova in the relatively nearby and exceptionally interesting galaxy Centaurus A and, as he commented, it 'stood out like an organ stop'. At once he notified the professional observatories, and spectra and magnitude measurements were being made in a matter of hours. This was just as well, because Evans had found something really out of the normal run.

Centaurus A is not a member of our Local Group of galaxies – a collection which includes our own system, the two Clouds of Magellan, the Andromeda and Triangulum Spirals, and a couple of dozen dwarf systems, plus the rather mysterious Maffei 1, about which we know little because it is so heavily obscured by dust in the plane of the Milky Way. The Local Group extends out to less than 7 million light-years, and Centaurus A is further off than that. Unfortunately it is too far south to be seen from Europe, but from southern countries a small telescope will show it, even to the curious dark 'lane' which stretches across it.

It was once believed that Centaurus A was made up of a merger of two galaxies which were colliding – that is to say, passing through each other. The individual stars would seldom hit each other, but would behave much in the manner of two orderly crowds passing in opposite directions. However, the dust and gas spread between the stars would be colliding all the time, so producing radio emission; and Centaurus A is a strong radio source. Later it was found that this explanation, attractive though it looked, would not fit the facts, because the amount of energy produced would be hopelessly inadequate to explain the intense radio emission. Centaurus A is a single galaxy, albeit an extremely peculiar one.

Evans' supernova, known officially as SN 1986G, seemed to lie in the dark dust lane (Fig. 63); so was it immersed in the dust, was it shining through, or was it on the near side? The first thing to do was to establish just what sort of supernova it was; and thanks to Evans' discovery, made before the star rose to its peak magnitude of about 7.5, this was easy. The supernova was definitely of Type I. It was also redder than it might have been expected to be, because it was veiled by the dust in the dark lane. Luckily it was not too far

149

NORTH

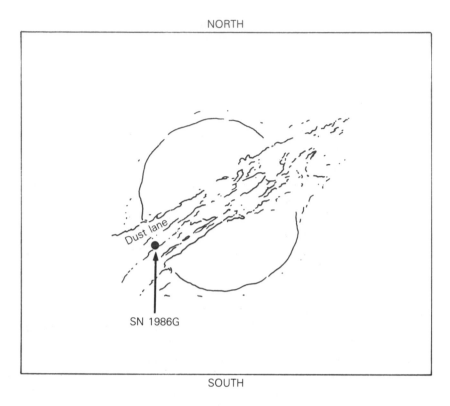

SOUTH

Fig. 63 SN 1986B position in Centaurus A, with the dust-lane, etc.

inside, or we would have been unable to see it at all. It was indeed what Paul Murdin of the Royal Greenwich Observatory called an 'optimum supernova'; it appeared at precisely the right moment in precisely the right place. Short of a supernova in the Local Group, or in the Milky Way galaxy, we could not hope for anything better.

The supernova has told us a great deal about the dust in Centaurus A, and effects in the spectrum also show that there are clouds of calcium lying between Centaurus A and our own Galaxy. Moreover, it indicates that Centaurus A itself is somewhat nearer than had been believed. The former estimate of its distance was 13 million light-years, but 1986G gives a value of more like 10 million light-years. This still places it beyond the Local Group; but if we have over-estimated its distance, the same error could apply to other galaxies too remote for their Cepheids to show up.

Would that there had been modern-type equipment at the time of the outburst of S Andromedæ – or that there had been an Evans or an Arbour to discover it, rather than a Hungarian baroness who did not know what it was or an assistant astronomer who was not allowed to make any announcement! But that is one of Nature's many quirks; and at least we may be sure that if a similar outburst occurs in a neighbouring galaxy, it will not be neglected in the way that S Andromedæ was just over a century ago.

The chance for which astronomers had been waiting came on the night of 23 February 1987. At the Las Campanas Observatory in Chile, the Canadian astronomer Ian Shelton had been taking long-exposure photographs of the Large Cloud of Magellan, that majestic star-system which is usually said to lie at a distance of 170,000 light-years (though astronomers at the Royal Greenwich Observatory prefer a lower value of 155,000 light-years). When Shelton developed his plate, he found to his astonishment that there was an unfamiliar bright spot near the famous Tarantula Nebula in the heart of the Cloud. Thinking that it might be a flaw in the plate, he went outside the dome and looked up at the Cloud direct. There, sure enough, was the bright spot – a supernova; the brightest since Kepler's Star of 1604.

Within a matter of hours, news of the discovery had been flashed all round the world, and every available telescope was directed toward the supernova. Observers in the northern part of the world yet again had to bemoan their bad luck! Studies from South America, South Africa, Australia and New Zealand were well under way as soon as the news broke; the old but trusty IUE, or International Ultra-violet Explorer satellite, was also directed to the star; so was an X-ray telescope, British-built but just launched by the Japanese. And Voyager 2, speeding its lonely way between the planets Uranus and Neptune, was also pressed into service.

Supernovæ may be expected to send out showers of those elusive particles known as neutrinos; and, sure enough, a neutrino burst was detected over a span of thirteen seconds on 23 February, three hours before the supernova was discovered. There were confirmatory observations from the neutrino detectors under Lake Erie in Canada, and Mount Elbrus in the USSR. So the timing was narrowed down to a few seconds; could we be watching the birth of a black hole, or at least a neutron star?

Some aspects of the supernova were puzzling. The characteristics varied between those of Types I and II, and there was a certain amount of disappointment that the magnitude never rose much above 4. But for the first time we have the chance of studying a supernova at relatively close quarters; it is an improvement on the Centaurus A star, at over 10 million light-years, or S Andromedæ, at over 2 million. Of course we are still desperately anxious to see a supernova in our own Galaxy; but if we cannot be satisfied in this respect, then we have to concede that an outburst in the Large Cloud of Magellan is certainly the next best thing.

XVII 3C-273: THE STAR THAT IS NOT A STAR

When Karl Jansky used his strange 'merry-go-round' aerial in 1931 to detect radio waves from the sky, he had no idea of the vital importance of the new branch of science which he was unwittingly creating. Neither, for that matter, did anyone else apart from an amateur named Grote Reber, who built the first 'dish' radio telescope just before the start of the war. Until the mid-1940s Reber was the only radio astronomer in the world. Today there are many thousands, but the progress of the infant science has not been without its hitches.

For example, it seemed logical to assume that the main radio sources would be bright stars. This was not correct. When the first discrete sources were identified, they were associated not with conspicuous objects, but with very faint ones. By 1944 Reber had tracked down radio waves from three constellations: Sagittarius, Cassiopeia and Cygnus. We now know that the source in Sagittarius comes from that mysterious region marking the centre of the Galaxy, where there may well be a black hole as well as a supernova remnant. The Cassiopeia source is undoubtedly the remnant of the supernova which flared up around 1670, but passed almost unseen because it was so heavily obscured by dust in the plane of the Milky Way. (John Flamsteed, the first Astronomer Royal, may have observed it and lettered it 3 Cassiopeiæ, under the impression that it was an ordinary star; the position fits fairly well, and the star should have been on view at the time of Flamsteed's observation, though today 3 Cassiopeiæ is absent from the sky.) That left the Cygnus source, which proved to be very different.

The main trouble with early radio telescopes was their poor resolution. Two sources which were quite widely separated in the sky would merge into one, and it was impossible to obtain precise positions for any of them, though some could be deduced – in

152

particular, the source in the Crab Nebula, known officially as Taurus A. Finally, in 1951, a new type of radio telescope was built, and the position of the Cygnus source (Cygnus A) was found satisfactorily enough for two astronomers at Palomar, Baade and Minkowski, to identify it optically. They used the 200-inch reflector, then the largest in the world, to show that Cygnus A coincided in position with a dim, peculiar-looking galaxy. When they obtained a spectrum, they had a major shock. All the lines were strongly red-shifted, which meant that the galaxy was racing away from us. This in itself was not unexpected; Edwin Hubble had already proved that every group of galaxies is receding from every other group, so that the entire universe is in a state of expansion. But Cygnus A was leaving us at 10,000 miles per second (16,000 km per sec.), which, if the red shifts were reliable guides, corresponded to a distance of 1000 million light-years.

Yet Cygnus A was one of the most powerful radio sources in the sky, so that from that distance its real output had to be immense. Astronomers began a systematic hunt for other sources like it, just in case Cygnus A might prove to be a cosmic freak – in which case it could lead to a totally erroneous distance-scale.

Gradually it became clear that some radio sources came from regions in the sky which appeared blank, or from what looked like dim, often bluish stars. One of these sources was listed as 3C–273, because it was the 273rd entry in the third catalogue of radio sources drawn up at Cambridge. It seemed to coincide in position with a star of magnitude 12.8, bright enough to be seen easily with a telescope of the size owned by the average amateur astronomer. The star itself looked delightfully innocent, and there were some doubts as to whether it could really be the culprit; it could so easily have been a normal, fairly nearby star which happened to lie in the foreground.

For once Nature decided to be helpful. 3C–273, much the brightest of these 'suspect' objects, happens to lie near the Zodiac, so that at times it can be hidden or occulted by the Moon. Occultations of stars are common enough; as the star is covered by the approaching lunar limb it snaps out as suddenly as a candle-flame in the wind, because a star is virtually a point source. Since the position of the star in the sky is known, the position of the Moon's limb at the moment of immersion is also known; and for many years the exact movements of the Moon were not determined as accurately as astronomers wanted, so that occultation observations were important. (To a certain extent, they still are.) The radio source 3C–273 was not likely to be a point source, but the occultation method could be expected to give at least a reasonably good idea of where it was in the sky.

An occultation was due in 1963. Nature, having offered to help, then made things more difficult by deciding that the occultation would not be visible from England, where the 250-foot 'dish' at Jodrell Bank was in full working order, or from the United States. However, it would be observable from Australia, and at Parkes, in New South Wales, there is a 210-foot 'dish' which was completed in 1961 and which remains one of the largest and best in the world. The Parkes team, made up of C. Hazard, M. Mackey and A. Shimmins, decided to try, but when they did some calculations they found that there were extra problems; 3C–273 would be so low over the horizon that the dish could not be tipped down far enough to reach it. Drastic action was called for. Local obstructions were ruthlessly hacked down, and part of the telescope's gearing mechanism was dismantled, with teeth being filed off. When everything possible had been done, the astronomers found that they would be able to record the very end of the occultation. They did so, and for the first time the position of a radio source of this type was found with real precision. It really did coincide with the starlike 3C–273.

Having played their part, the Parkes team restored their telescope to its original condition, and contacted Palomar. There, Maarten Schmidt used the 200-inch reflector to photograph the spectrum. When he analysed it, he realized to his amazement that 3C–273 was not a star at all, but something quite different. The spectrum was unlike that of any star, and the absorption lines of hydrogen were so tremendously red-shifted that they indicated a distance of thousands of millions of light-years. This seemed to make no sense at all. 3C–273 looked small; it had always been mistaken for a normal star in our Galaxy (photographs of it were found dating back as far as 1885); but if it were as remote as its spectrum showed, then it would have to be shining as brilliantly as several hundred galaxies put together. Assuming the average galaxy to contain 100,000 million stars, it takes an effort of the imagination to realize just what this means. 3C–273 became known as a QSO or Quasi-Stellar Object; today this is usually abbreviated to 'quasar'.

Once the breakthrough had been made, more and more quasars were found, some of them with much greater red shifts than 3C–273. A new line of investigation was opened up, though it had really started with the work of Hubble forty years earlier.

It was Hubble who had found short-period variable stars in the Andromeda Spiral, and used them to establish that the 'starry nebulæ' really are external galaxies. Later Baade had shown that the original Cepheid scale was wrong, and that the galaxies were twice as far away as Hubble had thought. This also explained why no RR Lyræ stars had been found in the spirals. They were there, sure enough, but because they are much less powerful than the Cepheids they were too faint to be seen at the increased distance.

It was already known that all the galaxies beyond our Local Group show red shifts in their spectra. This had originally been found by Vesto Melvin Slipher at the Lowell Observatory in Arizona, but had been interpreted by Hubble together with his colleague Milton Humason. (Humason had a strange beginning to his scientific career; his first post at the Mount Wilson Observatory was as a mule driver!) What Hubble and Humason found was that the farther away a galaxy lay, the greater its red shift, and therefore the greater its speed of recession. This led on to what we usually call the Hubble Constant, which links distance with velocity.*

The distances of the nearer galaxies can be found by using the Cepheids. Farther away, when we lose the Cepheids, we can make estimates by using the brightest supergiants, on the assumption that the supergiants in other galaxies are of the same mean luminosity as the supergiants in our own Milky Way system. Further out still we can use supernovæ, when Nature is co-operative enough to produce them. But eventually we lose even the supernovæ, and our only remaining method is to measure the red shifts and then use Hubble's relationship to calculate what distance this particular red shift means.

We are in some danger of arguing in a circle, because we are assuming that the red shifts in the spectra of galaxies (and quasars) are due entirely to the Doppler effect; if not, they will give misleading results. This is where the quasars become so important. They are much more powerful than galaxies, and can be seen out to what we believe to be greater distances. If they are as remote as we believe, what is their source of energy?

All manner of theories were proposed, some of them more plausible than others. One ever-present obstacle was the small size of the quasars. Some of them, including 3C–273, varied in light over short periods of a few months, and this proved that they really were relatively small, because the period of variation is limited by the amount of time taken for light to cross the object from one side to the other. Astronomers had to decide how so much power could be packed into an area smaller than the diameter of the Solar System.

One early idea involved chains of supernovæ. It was suggested that one supernova outburst would trigger off a second, a second would produce a third, and so on, in which case there would always be several supernovæ 'performing' simultaneously. This did not attract much support, partly because there seemed no conceivable way in which supernovæ could affect each other and partly because

---

*The value of the Hubble Constant is still a matter for debate. One estimate is 55 kilometres per second per megaparsec: that is to say, for each megaparsec of distance a galaxy recedes by an extra 55 kilometres per second (one megaparsec is equal to 3.26 million light-years).

the process would not be continuous in any case. Neither did the quasar spectra agree with any such interpretation.

Next came the proposal that the power came from a super-cluster of exceptionally luminous stars, but this seemed to involve a total light-emitting area much larger than a quasar. Then, in December 1964, Sir Fred Hoyle suggested that a quasar might be a 'white hole', the opposite of a black hole. Material disappears from the outside universe as soon as it passes over the event horizon of a black hole; if this material could gush up elsewhere it could, it was claimed, produce a white hole – a sort of cosmic geyser. But again the whole idea seemed far-fetched, and I doubt whether anyone supports it today.

Anti-matter, then? The great Swedish scientist, Hannes Alfvén, has proposed that there are two types of matter in the universe: our type (koinomatter) and its exact opposite (antimatter), so that when the two types meet they annihilate each other to produce energy. If a koinomatter galaxy met an antimatter galaxy, the mutual destruction would produce an outburst of quasar-like proportions. Yet Alfvén's antimatter has never been proved; there is no observational evidence in favour of it (it is not the same as what physicists normally call antimatter), and to be frank the whole theory seems so highly speculative that it ceases to be credible.

Obviously, then, astronomers turned back to their everlasting stand-by, the black hole. On this picture, a quasar would be the nucleus of a very active galaxy, deep inside which there would be a massive black hole – a cosmic cannibal, swallowing up material (including complete stars) so that the material about to be engulfed would send out the vast amount of energy we receive. Recent observations have not been conclusive, but we are now quite sure that quasars really are the cores of active galaxies, and in some cases the surrounding galaxies have been observed, though they are drowned by the much greater brightness of the cores. The black hole idea still has considerable support, and in any case there seems every reason to suppose that the quasars depend upon gravitational energy in some form or other.

It is true that some astronomers question the reliability of the red shifts in quasar spectra. Sir Fred Hoyle is one. He maintains that the red shifts are not pure Doppler effects, so that our distance estimates for the quasars are hopelessly wrong; in his view the quasars may be not far beyond the edge of our Galaxy, perhaps at distances of tens or even hundreds of millions of light-years, but certainly not thousands of millions. In America, Halton Arp, working at the observatories of Mount Wilson (California) and Las Campanas (Chile) has found a quasar-like object which seems to be joined to a nearby galaxy by a luminous bridge, presumably made up of gas intermingled with stars – and yet the red shifts of the quasar and the galaxy are completely different. If we follow the conventional explanation, the galaxy is receding at 1100 miles per second (1760 km per sec.) and the quasar at 13,000 miles per second (about 21,000 km per sec.), so that there could be no real connection between the two. Moreover, Arp has accumulated many photographs which show galaxies and quasars apparently lined up, yet with different red shifts. There are only two possible explanations. Either the alignments are due to sheer chance, or else the red shifts are tricking us.

It is fair to say that the latter a minority view, though it certainly cannot be disregarded. Meanwhile, let us follow the general trend of opinion, and consider that the quasars are as luminous and remote as their spectra indicate. How far out can we probe?

3C–273, which started it all, is the brightest of the quasars. Most of the others are much more distant. In April 1982 a systematic search by the Parkes observers led to the identification of a quasar known nowadays as PKS 2000–330, which had a higher red shift than any previously found; it was optically identified by observers using the 158-inch reflector at the Siding Spring Observatory, and was estimated to be receding at around 90 per cent of the velocity of light, which would mean a distance of about 13,000 million light-years. It did not hold the record for long. In 1986 photographs taken at Siding Spring with the telescope known as the 'UKS' (United Kingdom Schmidt) enabled two Cambridge astronomers, Stephen Warren and Paul Hewett, to track down an even more remote quasar. It lay in the southern part of Sculptor, and had a visual magnitude of below 19; but its spectrum could be studied, and the results seemed to be clear-cut. The recessional velocity is over 90 per cent that of light.

This discovery, incidentally, was made by optical methods; not all quasars are radio sources, and so far no radio emissions have been detected from the Sculptor object. They may be found eventually, but they are not particularly powerful even if they exist.

We may be approaching a 'point of no return'. If the law of increased velocity with increased distance holds good, then we must reach a distance where a quasar or a galaxy is moving away at the full speed of light – 186,000 miles per second (300,000 km) – in which case we will be unable to see it. This will mark the boundary of the observable universe, though not necessarily of the universe itself. The cut-off point seems to lie somewhere between 15,000 million and 20,000 million light-years, which, on the cosmic scale, is not too far beyond that of the Sculptor quasar.

When we look at an object far away across space, we are also looking backwards in time; we see the Sculptor quasar not as it is today, but as it used to be perhaps 14,000 million years ago, when the universe was young. But how large is the universe? Is it finite, or infinite? Did it begin with a 'big bang', when all matter came into existence at the same moment, or are we living in a universe which alternately expands and contracts, so that there is a new 'big bang' every 80,000 million years or so? These are questions which we cannot yet answer; as scientists, we are strong on detail, but woefully weak on fundamentals.

We are doing our best to find out as much as we can. Each year brings its quota of new theories and new developments, but in the whole of modern astronomy one of the most significant of all discoveries has been that of the nature of the quasars – beginning with 3C–273, the star that is not a star.

In this book I have tried to use some famous stars, and others which are not so famous, to demonstrate the ways in which astronomers of the past and present carry on their work. I hope I have interested you. Finally, let us never forget that we on Earth owe our existence to one star alone – our Sun – and that to us, if not to other beings elsewhere in the cosmos, the Sun is the most important star of all.

FURTHER READING

Clark, David H., *The Quest for SS433*, Adam Hilger, 1987.

Malin, D. and Murdin, P., *Colours of the Stars*, Cambridge University Press, 1984.

Moore, Patrick, *History of Astronomy*, Macdonald, 1984.

Moore, Patrick, *The New Atlas of the Universe*, Mitchell Beazley, 1986.

Murdin, L. and Murdin, P., *Supernovæ*, Cambridge University Press, 1986.

Norton, A.P., *Norton's Star Atlas*, Longman, 1986.

Petit, M., *Variable Stars*, Wiley, 1987.

Tucker, W. and Giacconi, R., *The X-Ray Universe*, Harvard University Press, 1985.

The periodical, *Astronomy Now* (Intra Press), contains up-to-date articles and information; it is published monthly from September 1987.

INDEX